Architectural Patterns and Techniques for Developing IoT Solutions

Build IoT applications using digital twins, gateways, rule engines, AI/ML integration, and related patterns

Jasbir Singh Dhaliwal

BIRMINGHAM—MUMBAI

Architectural Patterns and Techniques for Developing IoT Solutions

Group Product Manager: Preet Ahuja

Publishing Product Manager: Surbhi Suman

Senior Editor: Shruti Menon

Technical Editor: Nithik Cheruvakodan

Copy Editor: Safis Editing

Project Manager: Neil D'mello

Proofreader: Safis Editing

Indexer: Rekha Nair

Production Designer: Nilesh Mohite

Marketing Coordinator: Rohan Dobhal

Senior Marketing Coordinator: Linda Pearlson

First published: August 2023

Production reference: 1310823

Published by Packt Publishing Ltd.

Grosvenor House

11 St Paul's Square

Birmingham

B3 1RB.

ISBN 978-1-80324-549-2

www.packtpub.com

To my mother, Joginder Kaur, and to the memory of my father, Surjit Singh, for their immense sacrifices and teaching me the value of honest and sincere hard work. To my wife, Ramandeep Kaur, for being my loving partner and encouraging me to pursue my dreams, irrespective of the prevailing circumstances. To my lovely daughters, Kiranpreet Kaur and Manpreet Kaur, for ensuring that there is never a dull moment in our lives.

– Jasbir Singh Dhaliwal

Foreword

Computing has evolved from its humble roots in the 1940s, with an initial focus on the fundamentals of mathematical problems; to the 60s and 70s, where it was centered on symbolic systems, wherein the field first began to confront issues of complexity; to the 80s, with the rise of personal computing and the problems of human/computer interaction; then on to the 90s and our present century, addressing distributed and connected systems at scale. While these kinds of systems tend to dominate the narrative of computing, not every system is a Google, a Facebook, or an X, representing systems with soft edges and demanding elastic computing infrastructures at a global scale. There is another class of systems important to humanity; systems that touch and interact with the real world. Mind you, these kinds of systems have been present since the 40s with the advent of computers such as Whirlwind, but what is different is that now we see the intersection of cloud computing and the **Internet of Things** (**IoT**), with millions upon millions of sensors and actuators interfacing with the physical world.

This book is a comprehensive guide to building, deploying, and evolving software-intensive systems for IoT. You will find here solid, pragmatic advice on how to design such systems, how to evaluate them, and how to deliver them. Three things in particular delight me about this book: there is a clear emphasis on design patterns; broad coverage across problem domains, from manufacturing and agriculture to cities and beyond; and coverage of the connection of IoT systems to contemporary developments in artificial intelligence.

I found this to be a compelling, well-written, and very approachable book from which I learned some new things, and I hope you will too.

Grady Booch

ACM Fellow, IBM Fellow, IEEE Fellow, and IEEE Computing Pioneer

Contributors

About the author

Jasbir Singh Dhaliwal has over 26 years of software development and management experience, including 10 years in delivering complex IoT projects. Currently employed with IBM as a Principal Architect (IoT and cloud) and considered a thought leader with over 31 IoT patents, Jasbir has a deep understanding of IoT concepts/architectures, which is refined by delivering IoT projects in diverse domains such as consumer goods, smart buildings, healthcare, precision agriculture, automobile, and manufacturing. His extensive experience in both the public cloud and embedded domains gives him a unique edge in conceiving innovative end-to-end IoT solutions and applications. He holds a bachelor's degree in computer science and engineering from Punjab Engineering College, India.

About the reviewers

Amit Bahuguna has worked for over two decades in various software engineering roles, across different geographies, covering the entire software development life cycle, for both government and private enterprises. Amit has specialized in IT architecture over the last 15 years and has worked on the architectures of several operational technologies and IoT projects during this period. He received a B.E. in electronic and communications engineering from Manipal Institute of Technology, and a master's in software systems engineering from the University of Melbourne. He is currently employed as a Solution Architect with the ICT department of the University of Sydney. He is also a member and certified professional of the Australian Computer Society.

I am thankful to my family, friends, and colleagues for sharing their valuable time and experience in acquainting me with, teaching me about, and enriching my knowledge of ICT. I am grateful to the open communities where access to information and learning are unrestricted and people are interested in helping one another, and I am continually humbled by how much I don't know.

Atif Ali is a mechatronics engineer specializing in automation, robotics, and IIoT, with a focus on electro-mechanical design and software integration for complex physical systems and processes. He is a passionate open source advocate, focused on the intersection of hardware and software for perceptive physical world observability. Atif has been heavily involved in both industry and academia for the past several years, in sectors ranging from clean energy and web3 to cybersecurity and mining, holding positions at ETH Zürich, UBC Vancouver, and Grafana, among others. He also served as a technical advisor for various start-ups, and was on the faculty board for the Global Masters in IoT at the Zigurat Global Institute of Technology.

Bart Szczepanski, a co-founder of a Polish IoT firm named Rebels Software, assists customers with attentive care and guides them through the world of IoT. At Rebels Software, Bart ensures that customers receive the necessary support and expertise in navigating the complexities of IoT. He holds a master's degree in electronics and telecommunication from Wrocław University of Technology and takes pride in his roles as a dedicated husband and a father to three children.

Philipp Staeheli is an experienced IoT Solution Architect working for an international and innovation-driven cleantech company. He has a deep understanding of IoT from both technical and organizational/business aspects. Prior to his current role, he worked for several years as a hardware and software developer as well as a project leader across different industries. Philipp is passionate about creating customer value and sharing his knowledge with others. He holds a degree in electrical engineering as well as business administration and has obtained several certifications in software architecture, cybersecurity, and Agile methods.

Acknowledgments

I would like to thank the following people:

Akhilesh Gupta, Principal Architect, FEV India Pvt. Ltd. – automotive connectivity and connected vehicles, for helping me to differentiate between IoT and general cybersecurity. His guidance helped me understand the challenges in securing IoT devices, especially connected vehicles.

Amar Mundi, Principal Architect, HCL Technologies, is one of the best technical minds and is always a phone call away for any kind of support. His guidance was instrumental in shaping a few of the initial chapters.

Amarjit Singh Dhaliwal, Sant Longowal Institute of Engineering and Technology (SLIET), for introducing me to the relevant people in academia. His support has ensured that this book has a good balance of theoretical concepts as well as practical guidance.

Glen D. Singh from the Packt team, for helping a budding author like me to understand the nuances of content preparation and presentation and guiding me to take the book to the right audience.

Grady Booch, IBM Fellow, for providing consistent mentoring and encouragement. He is one of the most generous souls that I have encountered in my professional life. His encouragement is the sole reason that this book was able to see the light of day.

Harsha Vacher, founder of KTech Labs Inc, for providing insights into IoT security.

Jerry Cuomo, IBM Fellow, for providing guidance on how to accelerate the pace of content preparation while not compromising on overall quality.

Karnail Singh Sidhu and Narinder Sidhu, the owners of Kalala Wines, Canada, for sharing their perspectives on the nuances involved in wine/grape farming. Their knowledge of the agriculture domain is truly impressive.

Kavita Mittal, Application Security Specialist at Thoughtworks, for enlightening me on the differences between IoT and IT security. I am always impressed by her technical know-how and supportive nature.

Kim Bartkowski, IBM Distinguished Designer, for sparing time from a very busy schedule to give me tips on how to enhance the visual appeal of the diagrams.

Neetika Jain, IBM Delivery Manager, for being a constant guide and helping in fine-tuning my understanding of concepts related to analytics, security, and a host of other topics.

Nihar Kapadia from the Packt team, for reviewing the content in detail and providing constructive input.

Professor Kalsi, SLIET, Longowal, for guiding me on the nuances involved in smart agriculture and its significance in an Indian context.

Raghu Ramaswamy, IBM Executive IT Architect, for your constructive feedback and encouraging words.

Rohan Dobhal from the Packt team, for helping me understand the mechanics of efficient marketing and promotion.

Sameep Mehta, Distinguished Engineer at IBM, for sharing personal experiences of writing a book. His ability to churn out content in a short time inspired me to recalibrate my writing speed, ensuring that the manuscript was completed on time.

Sean Lobo and Neil D'mello from the Packt team, for keeping me on my toes and ensuring that all chapter submissions were done on time.

Shivani Aggarwal, Solution Architect, HCL Technologies, for sharing insights regarding how to provide security for constrained devices.

Shruti Menon from the Packt team, for the diligent editorial reviews and always going above and beyond to ensure that the content was crisp yet complete. Her attention to detail is exemplary.

Sonia Mezzetta, Program Director, Product Management, Data Fabric Solution Architecture at IBM, for sharing her experiences of publishing a book and practical tips for ensuring the book's wider circulation.

Surbhi Suman from the Packt team, for always being patient even when answering a list of long-winded queries. Her ability to get things done is amazing.

Ulrike Vauth, Distinguished Engineer at IBM, for regular mentoring sessions. Your guidance on how to balance professional and personal aspirations helped me to effectively juggle both worlds.

Last but not the least, special thanks to those who intended to help but were genuinely hard-pressed for time.

Table of Contents

Part 2: IoT Patterns in Action

4

5

6

7

8

Pattern Implementation in the Agriculture Domain   121

Part 3: Implementation Considerations

9

Sensor and Actuator Selection Guidelines   143

10

Analytics in the IoT Context   167

11

Security in the IoT Context   189

Part 4: Extending IoT Solutions

12

Exploring Synergies with Emerging Technologies   213

13

Epilogue   249

Index   269

Other Books You May Enjoy   280

Preface

The book helps you to apply modern architectural patterns and techniques to achieve scalability, resilience, and security in intelligent IoT solutions built for diverse domains such as manufacturing and industry, consumer goods, agriculture, and smart city applications.

Who this book is for

This book is for IoT systems and solutions architects, as well as other IoT practitioners such as developers, technical program and pre-sales managers, and so on, who are interested in understanding how various IoT architectural patterns and techniques can be applied for developing unique and diverse IoT applications.

What this book covers

Chapter 1, *Introduction to IoT Patterns*, provides basic knowledge about IoT concepts that will help in understanding the architectural patterns and use cases detailed in subsequent chapters.

Chapter 2, *IoT Patterns for Field Devices*, lists the architectural patterns that are relevant to field devices, including device gateways, digital twins, and device management.

Chapter 3, *IoT Patterns for the Central Server*, discusses the architectural patterns that are relevant to a central server, such as AI/ML integration, rule engines, file upload, and enterprise system integration.

Chapter 4, *Pattern Implementation in the Consumer Domain*, explores how the patterns covered in the previous chapters can be combined to realize use cases (home automation and smart egg boilers) in the consumer domain.

Chapter 5, *Pattern Implementation in the Smart City Domain*, offers insights into how architectural patterns can help in realizing use cases in the smart city domain, including smart speakers, condition monitoring for perishable goods, driver behavior monitoring, and the automatic replenishment of consumables.

Chapter 6, *Pattern Implementation in the Retail Domain*, explains how the patterns learned in the previous chapters can help in realizing use cases (real-time tracking in retail outlets) that are relevant to the retail domain. Also, the chapter lists the retail domain-specific concepts that are related to IoT solutions.

Chapter 7, Pattern Implementation in the Manufacturing Domain, starts with the required know-how about smart manufacturing and then details the implementation of a use case (the automatic inspection of finished goods) using IoT architectural patterns.

Chapter 8, Pattern Implementation in the Agriculture Domain, describes the benefits of integrating IoT with the agricultural domain and also provides details about the implementation of a specific use case – a land consolidation platform.

Chapter 9, Sensor and Actuator Selection Guidelines, provides details about key concepts related to sensors and actuators and outlines the guidelines for selecting the most appropriate sensor or actuator depending on the use case requirements and related constraints.

Chapter 10, Analytics in the IoT Context, presents details about how the ingested IoT data can be used to generate insights. The chapter focuses on analytics as it relates to IoT implementations.

Chapter 11, Security in the IoT Context, discusses the specific considerations that need to be taken to ensure that IoT solutions are completely secure.

Chapter 12, Exploring Synergies with Emerging Technologies, explores the potential of combining IoT with other emerging technologies (such as blockchain, generative AI, 3D printing, and AR/VR) to create more powerful applications/use cases.

Chapter 13, Epilogue, identifies the practical challenges that are typically encountered while implementing IoT solutions as well as specific tips for mitigating those challenges. It also lists the key learnings that the author had while working on IoT projects.

To get the most out of this book

Before reading this book, the reader should acquire basic know-how about IoT. Prior knowledge of IoT's fundamental concepts and its application areas is good to have before reading this book but is not mandatory.

Image credits

Several images in *Chapters 4, 6* to *10, 12*, and *13* have been created using assets from `freepik.com` and `flaticon.com`.

Conventions used

There are a number of text conventions used throughout this book.

Bold: Indicates a new term, an important word, or words that you see onscreen. Here is an example: "Devices such as video cameras send the data to a **Device Gateway** (**DG**) over protocols such as Wi-Fi."

> **Tips or important notes**
> Appear like this.

Get in touch

Feedback from our readers is always welcome.

General feedback: If you have questions about any aspect of this book, email us at `customercare@packtpub.com` and mention the book title in the subject of your message. You can also contact the author on LinkedIn (`https://www.linkedin.com/in/jasbir-singh-dhaliwal-617a193`) or via email (`jas_singh14@yahoo.com`).

Errata: Although we have taken every care to ensure the accuracy of our content, mistakes do happen. If you have found a mistake in this book, we would be grateful if you would report this to us. Please visit `www.packtpub.com/support/errata` and fill in the form.

Piracy: If you come across any illegal copies of our works in any form on the internet, we would be grateful if you would provide us with the location address or website name. Please contact us at `copyright@packt.com` with a link to the material.

If you are interested in becoming an author: If there is a topic that you have expertise in and you are interested in either writing or contributing to a book, please visit `authors.packtpub.com`.

Share Your Thoughts

Once you've read *Architectural Patterns and Techniques for Developing IoT Solutions*, we'd love to hear your thoughts! Scan the QR code below to go straight to the Amazon review page for this book and share your feedback.

https://packt.link/r/1803245492

Your review is important to us and the tech community and will help us make sure we're delivering excellent quality content.

Download a free PDF copy of this book

Thanks for purchasing this book!

Do you like to read on the go but are unable to carry your print books everywhere?

Is your eBook purchase not compatible with the device of your choice?

Don't worry, now with every Packt book you get a DRM-free PDF version of that book at no cost.

Read anywhere, any place, on any device. Search, copy, and paste code from your favorite technical books directly into your application.

The perks don't stop there, you can get exclusive access to discounts, newsletters, and great free content in your inbox daily.

Follow these simple steps to get the benefits:

1. Scan the QR code or visit the link below:

https://packt.link/free-ebook/9781803245492

2. Submit your proof of purchase
3. That's it! We'll send your free PDF and other benefits to your email directly

Part 1:
Understanding IoT Patterns

Readers will be made aware of the fundamental IoT patterns that they can use when architecting IoT solutions. Each chapter will start with details about a pattern, its significance, and the type of scenarios under which the pattern would be usable.

This part comprises the following chapters:

- *Chapter 1, Introduction to IoT Patterns*
- *Chapter 2, IoT Patterns for Field Devices*
- *Chapter 3, IoT Patterns for the Central Server*

1
Introduction to IoT Patterns

The **Internet of Things (IoT)** has gained significant traction in the recent past and this field is poised for exponential growth in the coming years. This growth will span all the major domains/industry verticals, including consumer, home, manufacturing, health, travel, and transportation. This book will provide a novel perspective to those who want to understand the fundamental IoT patterns and how these patterns can be mixed and matched to implement unique and diverse IoT applications.

This introductory chapter details the architectural considerations that you must bear in mind while designing IoT solutions. Architecting IoT solutions is challenging as there are additional complexities due to the physical hardware selection, complex integrations, and connectivity requirements involved. This chapter also serves as a foundation for the patterns that will be introduced in the subsequent chapters.

In this chapter, we will cover the following topics:

- An overview of IoT
- IoT reference architecture
- Unique requirements of IoT use cases
- Recommended architecture principles and considerations

An overview of IoT

IoT has generated a lot of interest recently and has moved from a purely academic pursuit to the point where real use cases are being realized. IoT implementation is inherently complex due to multiple and diverse technologies (embedded, cloud, edge, big data, **artificial intelligence (AI)**, **machine learning (ML)**, and so on) being involved and due to the range of deployment options that are available (constraint devices in the field to the almost unlimited availablity of compute and other resources in the cloud). IoT enables diverse use cases and spans multiple domains (home automation, healthcare, track-and-trace, connected vehicles, autonomous driving, and more).

The relevance of IoT is only going to increase in the coming years because of the following reasons:

- IoT use cases encompass physical and virtual worlds and as a result, interesting and rich use cases can be developed (compared to purely virtual/software systems such as word processors, ERP systems, and more). It can be said that the scope and variety of IoT use cases is only limited by a person's imagination.

- The immense potential of IoT has been validated by both academics and implementors. This can be attributed to the following reasons:

  - The increased capability and efficiency of hardware components with a continous decrease in cost (and size) in line with Moore's law. Efficient battery utilization by the current generation of hardware components has also reduced the hassles of frequent battery replacement.

  - The rise of commercial cloud providers (hyperscalers), which enables unlimited scalability in terms of compute, storage, complex analytics, high-speed data ingestion, and more. These characteristics are very well suited to the needs of IoT applications. The following are some of the services that are provided by public cloud providers/hyperscalers and that can be leveraged while developing IoT use cases:

    - Device management

    - Firmware updates

    - Edge management/analytics

    - Device/data security

    - Digital twins

    - IoT analytics

    - Data ingestion

    - Data visualizations

    - Data storage

    - Video analytics

    - Notifications

  - Ubiquitous and low-cost connectivity in the form of traditional connectivity options (for example, Wi-Fi, 3G, and 4G) as well as connectivity options such as LoRaWAN and NB-IoT. Technologies such as NB-IoT are especially useful for IoT as they support long-range connectivity and provide long battery life. The advent of 5G has further enlarged the scope of IoT use cases by providing high bandwidth and minimal latency.

  - The increased maturity of related technologies such as blockchain, robotics, AI/ML, energy harvesting, AR/VR, drones, social media, and more. These technologies enable IoT practitioners to augment IoT capabilities with the capabilities provided by these technologies to push the boundaries of innovation further and envisage non-conventional ideas.

- The increased adoption of mobile and wearable devices. These devices enable *anytime access* to IoT data and help control and monitor IoT devices in real time.

- Increased market competition, which forces enterprises to treat data as the fulcrum for *decision-making* as well as monetization opportunities. IoT also acts as the foundation for operationalizing additional revenue models such as service revenue over and above product sale revenue.

It is important to understand how IoT systems are different from non-IoT systems. A few of the key differentiators are as follows:

- Humans play a vital role in operating and managing most non-IoT (traditional IT/OT systems) systems, whereas IoT systems are designed to operate on their own or with minimal human intervention.

- IoT devices are constrained in terms of the compute, storage, or both, whereas most non-IoT applications are deployed on standard workstations where an ample amount of storage and compute is available.

- IoT applications, once deployed, are expected to last for years (10 to 15 years is the norm in the manufacturing industry) compared to non-IoT applications, where the shelf life is less (typical refresh cycles range between 3 to 5 years). Accordingly, IoT systems must be architected by balancing both current and long term needs.

- Considerable heterogeneity is observed in the selection of the hardware/software components as well as connectivity protocols. This is because there are different technologies to choose from, and for each technology choice, there are multiple implementations and offerings available from vendors. In comparison, there are fewer technology options possible in non-IoT systems.

- There are differences in the characteristics of the data that is generated in IoT and non-IoT systems. All the seven V's of big data (velocity, variety, variability, volume, veracity, visualization, and value) are high in IoT systems compared to non-IoT systems.

- Very few IoT systems operate in isolation and are generally integrated with other enterprise systems. In IoT systems, there is a need to integrate **Information Technology** (**IT**) and **Operational Technology** (**OT**), as well as hardware devices. This presents an entirely new level of integration complexity that is rarely seen in traditional systems.

- Security is important in any connected system, but it becomes much more important in IoT as the attacks can result in physical harm (industrial robots gone rogue) in addition to reputation/financial loss. Additionally, most IoT field devices are installed in vulnerable environments where they can easily be tampered with. Therefore, the attack surface in an IoT use case is much larger than that of a non-IoT use case.

These unique characteristics of IoT systems can be visualized in the following diagram:

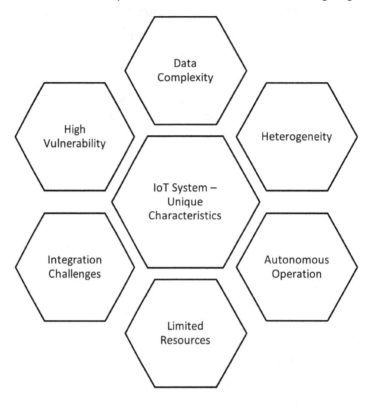

Figure 1.1 – Unique characteristics of an IoT system

This complexity can be quite daunting to anyone who has just ventured into the IoT domain. Although a rich variety of IoT use cases (or solution domains) is possible, there is also a certain degree of commonality that is present in most of the IoT use cases and related architectures. We have mentioned these similarities so that anyone who is new to this domain can understand the existing architectures and use cases.

IoT reference architecture

The **IoT reference architecture** follows a layered model, as shown in the following diagram:

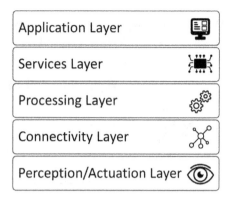

Figure 1.2 – Layered IoT reference architecture

Let's look at these layers in more detail:

- **Perception/actuation layer**: This layer indicates the physical layer where sensors (pressure, temperature, and so on) gather information about the environment. In turn, the environment is affected by the actuators (electric motor, thermostat control, and so on).

- **Connectivity layer**: This layer provides the connectivity required to send data (perception data from sensors and control commands to actuators, and so on) to/from the aggregation/processing layer. This layer is realized by leveraging connectivity options (5G, Wi-Fi, NB-IoT, LoRA, and so on). The decision to choose a specific connectivity option depends on various factors such as range and bandwidth.

- **Processing layer**: The processing layer ingests, analyzes, and stores data received from the connectivity layer. The processing can be performed either near the data source (edge computing) or in a private/public cloud. Data processing and storage elements, such as databases, data streaming engines, and AI/ML algorithms, form part of this layer.

- **Services layer**: This layer connects the *processing layer* to the application layer. Another way of looking at this layer is considering it as a set of APIs that can be consumed by the *application layer* to develop IoT applications such as smart homes, precision agriculture, smart manufacturing, and more.

- **Application layer**: This layer represents the applications that are to be used by end users. These applications are typically hosted at the edge or in the cloud (central server) and are consumed using mobile devices as mobile apps. Alternatively, they can be deployed on a web server and accessed using browsers.

The IoT patterns listed in subsequent chapters will align with the *IoT reference architecture* that we just discussed. Additionally, other important IoT topics listed in the latter part of the book (such as data analytics and IoT security) will also build upon the understanding of this concept.

The layered reference architecture provides various benefits, such as independent scalability of different layers and enhanced maintainability as change is restricted to specific layers. The IoT patterns will help you develop the required functionality at the specific layer in less time and in a reproducible fashion. The architectural patterns (detailed in subsequent chapters) that are relevant to the different layers of the reference architecture are shown in the following diagram:

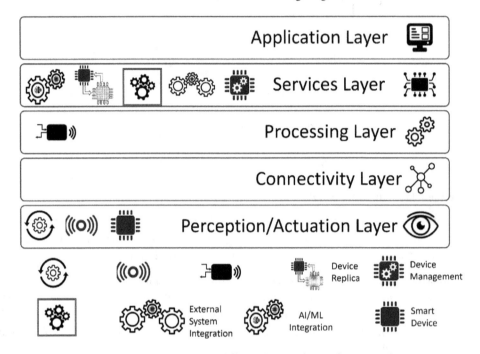

Figure 1.3 – IoT patterns realized at different layers of the reference architecture

Next, we will look at the unique requirements that we should be aware of while implementing IoT use cases.

Unique requirements of IoT use cases

IoT use cases tend to have very unique requirements concerning power consumption, bandwidth, analytics, and more. Additionally, the inherent complexity of IoT implementations (computationally challenged field devices on one end of the spectrum vis-à-vis almost infinite capacity of the cloud on the other) forces architects to make difficult architectural decisions and implementation choices. The diversity of the available implementation technologies and the absence of well-established standards are additional challenges that makes architecture decisions difficult.

This book attempts to alleviate some of the challenges associated with architecting IoT use cases by identifying the commonalities between the architectures that can support these use cases. It is important not to get blindsided by the diversity of the use cases and recognize the fact that diversity exists at the superficial level and *under the hood*. This book intends to bridge this gap in the current understanding by demonstrating how the implementation of diverse IoT use cases can be traced back to a handful of architectural patterns.

Before presenting the various IoT patterns, it is worth mentioning the unique expectations from IoT architectures that are different from non-IoT architectures:

- Sensing events and actuation commands have a wide range of latency expectations – from *real-time* to *fire and forget*.

- Data analysis results need to be reported/visualized/consumed on a variety of consumer devices – mobiles, desktops, tablets, and more. Similarly, data consumers have diverse backgrounds, data needs, and application roles (personas).

- One is often forced to integrate with legacy as well as cutting-edge devices and/or external systems – very few trivial use cases have isolated/standalone architectures. There is a considerable difference in the way the data is extracted from legacy versus non-legacy systems – legacy systems may internally collate the data and then push it to the external port (file transfer), whereas newer systems may push the data in a continuous stream (time-series data). This variability is one of the critical considerations when choosing a particular IoT architectural pattern.

- Varied deployment requirements – edge, on-premise, hybrid, the cloud, and more.

- Adherence to strict regulatory compliances, especially in medical and aeronautical domains.

- There are expectations considering immediate payback, **return on investment (ROI)**, business outcomes, and new service business models.

- Continuous innovation, which results in new services or offerings (especially by cloud vendors), forcing IoT architectures to be in *continuous sync mode* with these new offerings or services.

- The scarcity of skilled architects who can formulate *end-to-end* IoT solutions – although people with specific skills sets might be available (device architects, connectivity architects, and cloud architects); however, there are very few end-to-end IoT architects.

- No common standard for devices, device connectivity, IoT protocols, or the message transport layer leads to complex device management.

- Typically, IoT stacks don't operate in isolation and any non-trivial deployed IoT solution would need to integrate with other external systems (ERPs, AMDBs, MESs, and so on). Even here, there is no standard for how to integrate these systems seamlessly. The external systems typically predate IoT deployments by decades and are heavily customized with no consideration for integration needs.

- From one perspective, IoT implementation is a process automation initiative. In general, the process exists but is performed manually and IoT is expected to automate the process either partially or fully. Generally, these existing workflows are not documented and exist as part of tribal knowledge of the process practitioners, which poses challenges for IoT architects as they don't have clarity regarding the processes and workflows. Hence, they face a dilemma regarding which subprocesses should be automated to maximize their ROI – they have to decide if they are content with minor improvements (local optimization) and forgo benefits that can be accrued by considering global optimizations.

- Device life cycle management is a challenge in domains such as cardio medical devices as they can't afford downtime but still need a timely firmware update (especially patches related to security fixes, which can't be deferred beyond a certain point).

- The need to calibrate field sensors at regular intervals poses a challenge. The rate of drift varies from sensor to sensor and from one environment to another. There is a tendency to compensate for this drift by applying AI/ML models at the edge or in the cloud, but these steps are far from ideal as they lack accuracy and may not fully consider local or ambient conditions.

- Use cases that rely on positional information tend to have limited acceptance as all the locational sensors (indoor or outdoor) have limited accuracy.

- The migration of massive amounts of edge-processed historical data (accumulated over decades) to the cloud is another key architectural challenge that is observed in many **Machine-to-Machine (M2M)** to IoT transformation initiatives.

- The desired **non-functional requirement** (**NFR**) (scalability, availability, security, data residency/privacy, and so on) values vary from use case to use case and add another layer of complexity.

- Consumers of IoT data have diverse backgrounds (for example, the information needs of a *home automation* user would differ widely from an industrial user who wants to monitor *plant uptime*, which, in turn, would be different from the needs of paramedical staff using IoT for automated clinical trials), so they have different ways of operating and leveraging IoT systems. Although this may seem to have more bearing on device UI design, it can impact the solution architecture in subtle ways as well.

In the next section, we will list the architectural principles or considerations that will help you address the unique requirements of implementing IoT solutions.

Recommended architecture principles and considerations

Certain principles, which ensure that architectures, once realized, are scalable, modifiable, robust, and fault-tolerant are especially relevant for IoT architectures. Let's take a look at some of these:

- **Built on open communication protocols to support diverse device communication needs**: As IoT is an amalgamation of real (hardware) and virtual (software) realms, each of which evolves at its own independent pace. Robust IoT architectures should be flexible enough to support current and possible future enhancements in both these realms – for example, on the one hand, continual advancements are made for connectivity/power capabilities on the device/hardware side, while on the other hand, there are central server side advances regarding analytics and AI/ML capabilities. Hence, there is an inherent impedance mismatch between real and virtual worlds (concerning the rate as well as nature of these enhancements). IoT architects should not only be aware of this mismatch but should also incorporate the required considerations to support the use case requirements for a longer time frame. These requirements are partially handled by adhering to a layered architecture whereby the components in a specific layer can be *plugged in* or *plugged out* with minimal impact on the overall architecture.

- **Designed for "end-to-end" security**: Security is an important consideration for any software system, especially in cases where data or commands are communicated over public communication channels. However, in terms of IoT, security requires deeper consideration, primarily due to two reasons:

  - **Actions initiated in the real/physical world can't be rescinded, unlike the actions in the virtual/software world**: An irrigation pump that is instructed (maliciously) to start pumping water in an agriculture field would have pumped considerable water before someone detects the anomaly and initiates corrective action. This contrasts with the scenario in the software world, where a simple *update* instruction is sufficient to *undo/roll back* database changes. Scenarios can be even more disastrous in domains such as healthcare, where IoT systems often control human life (for example, an oxygen ventilator controlled by an IoT system).

  - **The attack vector is considerably broader compared to pure software systems**: This is because the complete data pipeline (end device > gateway > communication channel > central server > application) needs to be secured and each entity in the data pipeline has diverse applicable security requirements – end devices (with their inherently constrained compute/storage capabilities) can't support the security rigor that the central server can support, so each component's security vulnerabilities and the relevant security guardrails need to be independently analyzed. Similarly, data should be protected in transit as well as at rest at all times.

- **Enterprise integration enabled by the "API-first" approach**: Any production-grade IoT system will typically be integrated with other external systems to deliver full value. Real-world data collated by IoT systems is fed (*data push*) into external systems to enable richer use cases. Similarly, data from the external systems (*data pull*) is used to enrich the collated data. This type of integration is not possible unless the IoT system has been architected with *API-first* as one of the core architectural tenants whereby IoT data can be consumed by enterprise applications. These APIs also enable workflows that span both IoT and non-IoT (that is, external systems).

- **Satisfy diverse data needs**: IoT systems are leveraged by a diverse set of users, each with different backgrounds and information needs. Accordingly, it is important to capture the raw data needs of all the (current and future) stakeholders and to present the data in a way that is easily assimilable by a diverse set of stakeholders (personas). **Role-based access control** (RBAC) is one mechanism that shows the required information to stakeholders while at the same time obscuring non-relevant information. Also, some of the stakeholders will have real-time data needs (operators who want real-time notifications for emergency alarms), whereas others will want insights from consolidated data (batch processing). *Decoupling data ingestion from data processing* is one such principle that enables us to accomplish this need. Some of the other data collation/manipulation requirements are listed as follows:

- **Diverse (structured, semi-structured, and unstructured) operational data** from sources such as **Manufacturing Execution Systems (MESs)** and **Laboratory Information Management Systems (LIMSs)** should be consolidated in a common data store (data lake) either at the edge, the cloud, or both.

- **Separating streaming, batch, and right-time data** pipelines for scalability, efficiency, and cost optimization considerations. De-coupling data producers from consumers ensures a robust architecture as well as the flexibility of technology and implementation choices.

- **Technology-neutral architecture providing deployment flexibility**: IoT systems can be deployed in different configurations, such as on-premise, public cloud, private cloud, and/or hybrid multi-cloud configurations, depending on the customer's sensitivity to security as well as governance and regulatory needs. Considering this, the architecture should be generic enough that it can cater to diverse deployment needs and can be supported by multiple technology stacks. This is generally achieved by creating an IoT reference architecture (devoid of specific technology choices) and then transitioning to a technical architecture (where generic architectural components are replaced by specific technology components).

- **Design for high availability**: Although the need for high availability varies widely from one IoT use case to another, some use cases are categorized as *mission-critical* with almost zero downtime expectations, whereas others can accommodate a considerable downtime period. The central server architecture should mimic the uptime expectations as typically, less downtime translates into higher costs. In the context of IoT, high availability must be considered from an overall system perspective. For example, in scenarios where longer central server downtime is acceptable, end devices need to have higher data buffering capabilities (that is, greater storage space) to minimize data loss.

- **Support for "unlimited scalability"**: IoT deployments start small with a few end devices but tend to scale to a large number in a short duration. As a result, generally, in IoT solutions, *horizontal scalability* is preferred over *vertical scalability*.

- **Device communication considerations**: Data is communicated over a bi-directional communication channel between the gateway and central server. This channel can be supported by multiple communication technologies (with some of the common ones being cellular, Wi-Fi, LoRa, and SigFox). Considerations such as range (physical distance from the central server), payload size, battery life, and ambient noise play a role in finalizing the ideal communication technology for a particular IoT implementation. Some of the other considerations from the device side include the ability to store/buffer data in case of connectivity loss to central server, sleep/wakeup logic for conserving battery power, and data aggregation/filtering needs.

The following diagram summarizes the key architectural principles/considerations discussed in this section:

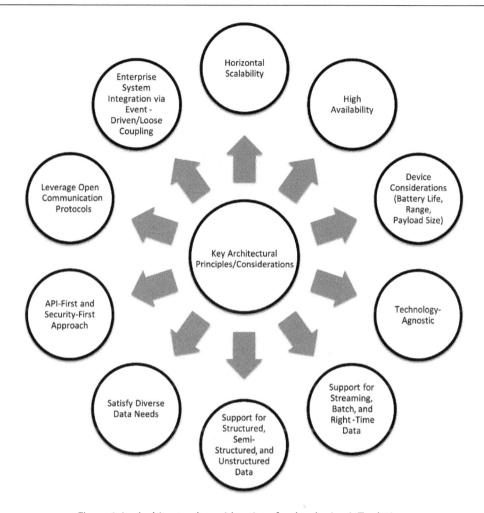

Figure 1.4 – Architectural considerations for developing IoT solutions

Summary

This introductory chapter helped you understand the architectural considerations need to be considered while developing or deploying IoT solutions. Additionally, the chapter provided contextual knowledge that will help you understand the patterns listed in this book. The characteristics that make IoT solutions different from other traditional software systems or IT solutions were discussed, along with information about the different layers of the IoT reference architecture. In the next two chapters, we will dive deep into the IoT architectural patterns.

2
IoT Patterns for Field Devices

This chapter lists the key patterns that are relevant to field devices or *Things*. After reading this chapter, you will be able to identify the existence of these patterns in IoT architectures. It provides details regarding the scenarios in which the patterns are suitable or applicable, along with the constraints that need to be considered. This will help you understand existing IoT architectures with relative ease.

This chapter covers the following three key patterns:

- **Device gateway (DG)**: A DG serves as a bridge between field devices (sensors, actuators, and so on) and the central server. In a standalone deployment (without the central server), DG coordinates actions between local devices (sensors and actuators).

- **Digital twin (DT)**: DT is used to maintain a virtual state of the field devices on the central server, thereby allowing for remote monitoring and control operations. By performing the required processing on the accumulated data, DT makes it possible to predict the future state of the field devices. In addition, DT helps to overcome intermittent connectivity isues.

- **Device management**: Device management helps configure, update, and manage the field devices and is hosted on the central server.

Let's look at these patterns in further detail.

Device gateway

DG is an important pattern as it helps link physical and virtual worlds. The physical world is monitored by sensors and actions are initiated by actuators, as per the commands that are sent by a DG. The notation that is used for the DG in this book is shown in the following diagram:

Figure 2.1 – Notation for the DG pattern

Important Note

The DG is also referred to as the *Field Gateway* in the IoT literature.

In addition to its role in enabling *edge/local* intelligence by hosting a **Local Rule Engine** (**LRE**) and performing latency-sensitive decisions, DG enables data communication with the central server, where more complex decisions (those requiring a global context) must be made. The need for DG arises because most sensors/actuators are constrained in terms of compute, memory, storage, or power, so they can't establish connectivity with the central server. A good practical example of DG is a smartphone as it connects to multiple devices (for example, headphones, lights over BLE, and so on) and sends the data over HTTP/MQTT to the central server. DG is functionally superior to routers as it can execute business logic at the edge rather than just routing traffic.

Another perspective is that the DG pattern encapsulates the different protocols or data formats in which sensors/actuators typically communicate. As there is no standardization of either the communication protocol (BLE, Wi-Fi, ZigBee, OPC UA, and so on) or data format across different sensors/actuators, DG fills the role of the *protocol translator*; it communicates on different communication protocols with sensors/actuators on one side and communicates over uniform data or communication protocols with the central server on the other, as shown in the following diagram:

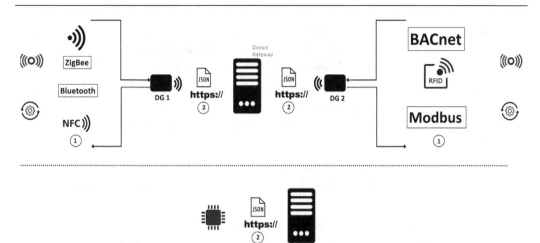

Figure 2.2 – A DG is capable of supporting multiple protocols; a comparison with a smart device

> **Important Note**
>
> DG acts as a connectivity enabler and protocol translator, and provides data-buffering capabilities. However, it may not be needed in scenarios where devices are *smart enough* to provide these capabilities.

At the top-right of the preceding diagram, the key functionality of the DG pattern is highlighted. As you can see, sensors, actuators, and other devices can interact with DG using a myriad of communication technologies/protocols (both wired and wireless, such as Wi-Fi, ZigBee, BACnet, Modbus, Bluetooth/BLE, RFID, NFC, and more; these are marked as **1** in the diagram). However, the interface between DG and the central server is of a single type (the diagram shows the interface as JSON over HTTPS, though other interfaces, such as AMQP and MQTT, are also used in the IoT ecosystem; they are marked as **2**). Another key distinguishing factor between **1** and **2** is that **1** is non-IP-based communication, whereas **2** is IP-based communication.

If the devices are smart (they have an adequate amount of compute, memory, and storage and can establish connectivity with the central server), DG is not required as the devices themselves are capable of connectivity and managing how data is transferred.

In addition to allowing *dumb* devices to send/receive data to/from the central server and acting as a protocol translator, DG provides additional functionalities as described in the following points:

- **Data aggregation/filtering**: In some scenarios, it is not required to send all the data that's been captured by the sensors to the central server (due to bandwidth constraints or the application not requiring data to be pushed at a high frequency). In this case, DG will accumulate data and send summarized data to the central server (from the past hour, sending data only if it is different from when it was read prior, and so on).

- **Data security at rest and in motion**: DG not only ensures that data that is stored locally is secured (that is, encrypted) but also that the data sent to the central server is secured by leveraging all the best practices related to **Authentication, Authorization, and Accounting (AAA)**.

- **Support local data access requirements**: DG allows you to access data locally (to remove dependencies from the central server for critical data access requirements) via APIs. In some scenarios, DG also hosts **Human Machine Interfaces** (**HMIs**) for visualization and reporting purposes.

- **LRE Support**: DG can enable LRE (the LRE pattern will be covered in *Chapter 3*, *IoT Patterns for the Central Server*), where generated events (for example, from sensors) are observed and suitable actions are triggered.

- **Firmware/configuration upgrades of connected devices**: Firmware upgrades for connected devices (or for DG itself) are pushed by the central server as needed. Similarly, configuration settings (for example, changes in the data capture frequency) are sent by the central server to DG. Commands may also to sent to DG for troubleshooting/diagnostics purposes.

- **Data buffering**: In the case of intermittent connectivity with the central server, DG can buffer data (depending upon the limit imposed by the available local storage) and send it once connectivity has been established, thus avoiding loss of data.

- **Application middleware for DG-hosted applications**: DG exposes APIs to report local analytics results (based on historical data), as well as current data that's been captured by sensors. Also, in certain critical scenarios, commands for actuators may be issued locally rather than you having to wait for the central server to make decisions. Again, this is made possible by the exposed APIs that are leveraged by applications hosted on DG.

Pattern summary

Let's take a look at the pattern summary for a DG:

- **Problem solved**:

  - **Business**:

    - Interoperability of diverse sensors/actuators that have different connectivity protocols with the central server

    - Ensuring the security of data at rest as well as during transmission

    - Enables real-time decision-making at the edge while avoiding a long (and non-deterministic) round trip to the central server

    - Prevents the loss of data in case of intermittent connectivity with the backend server by buffering data at the edge

    - Provides support for local/edge applications

- Reduces code duplicity as common functionalities (such as connectivity, data buffering, data encryption, and so on) are handled by one entity, such as DG, rather than by all the end devices (that is, sensors and actuators)

- Facilitates remote firmware upgrades as well as remote configuration updates of end devices, thus providing cost and effort efficiencies

- Enables connectivity for legacy end devices

- Enables efficient utilization of bandwidth by mechanisms such as data aggregation and data filtering

- **Technical**:

  - Provides *separation of concerns* as common functionalities (such as connectivity, data buffering, data encryption, and so on) are handled by the DG; the sensors and actuators handle interaction with the physical environment

  - Discourages tight coupling between DG and the *central server*, ensuring parallel development/enhancement of DG and the *central server*

- **Usage context**:

  - To provide connectivity to constrained/legacy devices

- **Example/usage scenarios**:

  - A smartphone acts as a DG to send data from end devices (for example, smartwatches and home automation controls such as thermostats, blinds, and so on) to the central server for complex analytics

  - In the manufacturing or industrial context, DG acts as a connectivity enabler to connect legacy machines to the central server for complex analytics (predictive maintenance, OEE calculations, and more)

- **Pattern rationale**:

  - To implement a functionality (such as central server connectivity, data buffering, data security, and so on) that is common for all the end devices, thus eliminating redundancy and complexity of the code/logic

  - To provide a uniform communication channel or protocol between diverse end devices (that have a variety of communication patterns and protocols) and the central server

  - To ease the development of local or edge level applications

  - To enable local and faster decision-making, especially in scenarios where network latency makes a round trip to the central server impractical

- To enable firmware or configuration updates for end (constrained) devices

- **Related patterns**:

  - Rule engine

- **Assumptions**:

  - The DG, like any other embedded system, will adhere to the best practices related to embedded systems security (for example, secure boot, data encryption, least privilege, role-based access control, and more)

  - The central server's connectivity endpoint would be preconfigured in the DG to enable connectivity at bootup time

  - To ease the development of applications on top of DG, the underlying (core) functionality of DGs (data buffering, data communication, and so on) should be exposed via a well-documented set of APIs

- **Considerations**:

  - Depending on the requirements of the use case, DG can be powered by a mains supply, or it can be battery-powered

  - To remove the dependency that DG applications have on underlying operating systems, applications can be deployed using container frameworks

  - To ease integrations with diverse end devices, it is recommended that DG implement a **Hardware Abstraction Layer** (**HAL**) to abstract hardware-specific nuances from other DG modules

  - To support local HMIs, DG generally provides one of the output interfaces as a video interface (VGA, HDMI, and so on). Similarly, a debugging port is provided to emit debugging/log messaging for local troubleshooting

  - In some specific scenarios, DG doesn't always need to be powered and needs to be *wakened* by special commands from the central server or local devices

  - As inbuilt storage for DGs is generally limited, DGs are typically equipped with external storage (an SD card, external hard drive, and so on) to accomodate extended periods of connectivity loss

  - Depending on the operating conditions (for example, extreme climatic conditions and/or dusty environments) in which DG must be deployed, it is housed in a rugged enclosure

  - In some domains, regulatory requirements may require that DG is compliant with standards such as the GDPR, CE/FCC/IC, and others

- **Anti-pattern scenarios**:

  - Not applicable in scenarios where end devices are capable (smart enough) of communicating directly with a backend server over IP communication (for example, a smartwatch with cellular connectivity to another smartwatch where it needs to latch onto a smartphone (DG) to send data to the backend server)

  - Not required in scenarios where functionalities such as protocol conversion, data aggregation, filtering, and buffering are not required and can be handled by the central server (in scenarios where bandwidth constraints are not present, for example)

This section covered the detailed specifications of the DG pattern. Now, let's look at another interesting device pattern, that is, DT.

Digital twin

A DT is a *virtual copy* of an IoT device that's deployed in the field. The concept is very similar to the process of creating a model (simulation) of a physical entity or process to understand its exhibited behavior. The notation for DT used in this book is shown in the following figure:

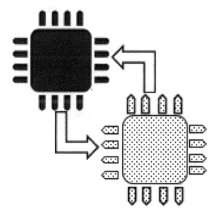

Figure 2.3 – Notation for the DT pattern

DT is an important pattern in the context of IoT as data may be transferred between the central server and field devices over unreliable communication channels (intermittent connectivity). Also, field devices may choose to *sleep* and only *wake up* during specific times to conserve energy. So, DT is also used to abstract users from the current state of field devices (wake up, sleep, and so on) and the communication channel's nuances.

DT also provides an encapsulation mechanism whereby users can view and set the state without being concerned with the actual state of the field devices and/or connectivity constraints. Seen in this manner, DT decouples the device's actual state from its virtual representation.

One way of defining DT is by considering it as a *living model* of a physical entity. This definition, in fact, lists two separate characteristics:

- **Living**: This characteristic indicates that the model of the physical entity is constantly replenished with changes that correspond to the physical/actual properties of the physical entities. As a result, the scenario where physical entities are not updating their states in a timely fashion (this can be due to multiple reasons, including to conserve the battery power of field devices) shouldn't be considered *true DT*.

- **Model**: This characteristic is DT's ability to effectively model a physical entity. DT can help you understand the physical entity (or physical phenomenon) in its entirety. As is the case for any software system, there is no *perfect architecture*; whatever model (or diagram) helps us understand the nuances of the physical process or environment can be termed as the *model* of the software system.

DT shows how the device under consideration behaves to its external stimulus, as well as how the device's internal parts interact with each other to exhibit the observed behavior. To cater to this expectation, the device must communicate its current state to the DT without significant time delays to the central server. From this perspective, DT is a mechanism to view and update an IoT device's state from a remote location. DT *mirrors* the device state on one hand and allows you to *remotely control* the device on the other.

However, when the device and its twin are separated in space by a reasonable physical distance, some amount of lag is unavoidable. Considering the reference architecture and the fact that DT is generally hosted in the cloud (or in private data centers) these delays are observed in the communication layer.

Different architectures can be envisaged to realize this pattern and their suitability can be measured/ compared based on how timely and accurately DT can project the state of the device(s). The time it takes for DT to process this should ideally be minimal and deterministic. Additionally, the more parameters that are being synced between the device and DT, the more closely it will replicate the device's status and behavior.

Any input data from the entity that's being monitored is sent to DT. After processing the data, DT sends back a command control, as shown in the following diagram, where data from a conveyor belt in a manufacturing plant is being fed into DT:

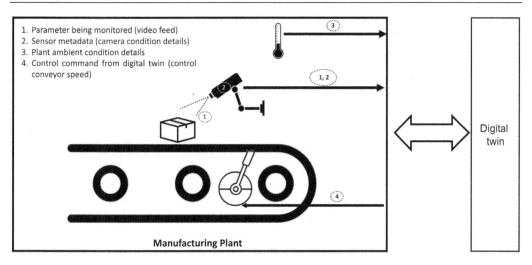

1. Parameter being monitored (video feed)
2. Sensor metadata (camera condition details)
3. Plant ambient condition details
4. Control command from digital twin (control conveyor speed)

Manufacturing Plant

Digital twin

Figure 2.4 – Data/control command interplay between the device and DT

As shown in the preceding diagram, the input data can be further classified as follows:

1. Data related to the entity being monitored – that is, the video feed of the part being manufactured. To avoid hogging the communication channel's bandwidth, the video feed is analyzed on the edge and only relevant notifications (events of interest) are sent to DT.

2. Sensor metadata (the condition of the camera, usage/operational characteristics, and so on).

3. The plant's ambient conditions (temperature, humidity values indicating the operating conditions where the part is being manufactured, and so on).

4. After analyzing the input data, DT sends a control/actuation command to the conveyor belt (used to start/stop the conveyor belt or to alter ambient conditions) to bring the manufacturing output within the desired tolerance levels.

As you can see, the device itself can be a complex state machine with numerous subcomponents that have diverse inter and intra-relationships between them. These subcomponents and their relationships need to be replicated in DT to *visualize* the device in its entirety. One benefit of DT is that it allows you to view the device's state at various levels of abstraction by zooming in and zooming out on the device properties, subcomponents, relationships, and more. The relationships between the subcomponents may change based on environmental factors and the same needs to be reflected in the DT. As a result, the device, in addition to posting the state to DT regularly, may need to post metadata related to the entity's topology/relationships.

In addition to the *abstraction scale*, DT should also incorporate a *time scale*, which allows the user to go back (to a past state) and forth (to a predicted or extrapolated state) as per the user's needs. Depending on the device's complexity and the desired level of accuracy, this may require simple regression/extrapolation or complex AI/ML models to be built.

Organizations can have different maturity levels of DT implementation. DT implementation is a journey and organizations can start small with a basic implementation; with time, DT can evolve into a full implementation. The following diagram provides a practical roadmap that organizations can use to evaluate their current and target DT maturity levels:

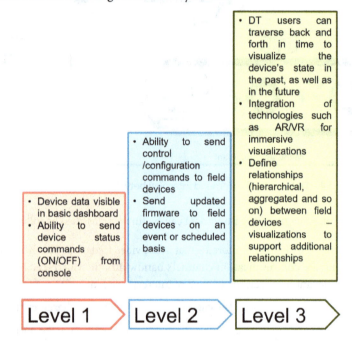

Figure 2.5 – DT's implementation maturity level for defining/tracking current/target maturity levels

There are additional expectations that all DTs should fulfill:

- Device data transmission might be interrupted due to connectivity failure with DT. In this case, DT should be able to simulate data for the period when data is not received (with an indication that the data is simulated and not real). Similarly, once connectivity has been established and data transmission resumes, the simulated data needs to be replaced with actual data.

- DT should be intelligent enough to understand whether the actuation command needs to be pushed to the device once connectivity has been re-established. This is because the situational context will have changed during the period connectivity was not available. Similarly, actuation commands will need to be merged or coalesced to remove redundant actuation (there's no need to push the same button multiple times, for example).

- Typically, DT won't exist in isolation and will be augmented with data from other auxiliary or external systems. Typically, data that's fetched from these devices are related to the metadata of field devices (for example, the installation date, aging data, performance characteristics, and so on). Another set of data can be obtained from other external systems to correlate and/

or validate the data that's been reported by field devices. For example, data from the weather website (external system) can be used to predict whether there is a need for fields to be irrigated, even though the fields' soil sensors report humidity levels that are lower than normal.

- DT should help visualize not only the current state of the monitored devices but also their past (historical) state and the future expected (extrapolated) state. Accordingly, the DT user should be able to perform a past or future *time travel* in terms of device condition or status. Similarly, the user should be able to *scale up* and *scale down* the abstraction level at which they want to monitor the state of the device. *Scaling up* would involve more aggregated or coarse readings/values/statuses while *scaling down* would indicate the need to get more raw or fine-grained readings.

- As in real life, the relationships between different entities can vary; for example, hierarchical, one-to-one, one-to-many, and so on. Taking the manufacturing industry as an example, machines are part of a department, such as the one responsible for painting, and these departments are part of the larger organization. Generally, policies that are applied at a higher level are expected to be automatically applied at lower levels, thus reducing management and compliance efforts. Similarly, the statuses or conditional metrics of the lower levels are automatically *rolled up* to the higher levels.

- In some scenarios, DT can also initiate ad hoc/need-based queries to understand and correlate observed data from physical entities. This would entail a complex sequence of sensing and actuation commands. An example is the DT of an oil rig, where some anomaly has been observed in oil production. Here, DT would initiate a visual inspection (using a fleet of camera-equipped drones) of the pipelines to understand and report the actual cause of the anomaly. Reducing false alarms is one benefit that justifies the extra processing and infrastructure.

By leveraging technologies such as AI/ML and in addition to traditional mathematical tools/methodologies such as interpolation, extrapolation, regression, and others, DT can be taken to the next level, as shown in the following examples:

- In precision agriculture/smart agriculture, the growth of the crops can be monitored more effectively by comparing them to the expected growth rate (real growth versus the expected growth). Another scenario could be comparing the growth rates of the crops of different farmers that are grown under similar environmental conditions.

- DT can play a crucial factor in monitoring and augmenting human wellbeing by tracking leading indicators such as blood sugar level, blood pressure, and others. Advanced analytical and heuristics techniques can provide valuable insights related to aging, future health, expected life span, and more.

- DT can help create self-healing systems where the appropriate instructions are sent to machinery to correct faults that have already happened or are expected to happen in the future.

> **Important Note**
> The functionality provided by DT will depend on factors such as the envisaged use case, hardware/software capabilities, expected latency, and available bandwidth.

Pattern summary

The pattern summary for a DT is as follows:

- **Problem solved**:

  - **Business**:

    - Accurately plan, with simulated or live sensor data, scenarios based on digital assets and operations

    - Understand the behavior of assets and operations by simulating stress conditions on DTs

    - Predict downtimes and breakdowns ahead of time

    - Control the functionality of assets and processes remotely with the desired state of assets and processes

    - Support semi- and fully autonomous operations

    - Provide a platform for analyzing data from field devices and initiating device actions

    - View historical states of field devices, as well as possible future or predicted states

    - Provide an abstraction of the device's status and a uniform interface for setting the state of the devices

    - View and set the values of the field devices from a central location

    - Simulate the behavior of a system to understand the potential gains and performance or efficiency issues before investing in the actual system implementation

    - Analyze a product, along with its operational context, to identify product refinement opportunities

    - Monitor the current and/or historical state of field devices in a consolidated view

  - **Technical**:

    - Advanced (3D) visualization of assets and processes

    - Extendable DT data model to cover multiple assets and processes

    - Extend the DT to a digital thread that spans beyond the organization

    - Provide context to IoT sensor data and consume the data to understand the behavior of the assets and processes

- Agile response to the changing current/reported state of assets and processes with a conditional simulation of the DT

- Eliminate the need to have continuous connectivity between devices and the central server

- Provide remote configuration of field devices and push firmware updates

- **Usage context**:

  - Understand and predict the behavior of the assets and machines

  - Control assets and processes remotely by sychronizing desired state and reported state. Typically, DT would be hosted/deployed on the central server

  - Integration with an AI/ML component to determine predictive values

  - Integration with external systems for data enrichment and to respond to *what-if* queries

  - DT would need to support additional devices and scenarios as the system evolves. Following a microservices-based architecture is advisable for circumventing potential scalability issues

  - Use DT to expose the functionality in the form of APIs as diverse consumers (VR headsets, mobile devices, web applications, and more) need to access information

  - Authentication, authorization, and **role-based access control** (**RBAC**) are required to support multiple consumer roles that have varied information needs

- **Example/usage scenarios**:

  - Understand the behavior of the building/facility before construction is complete, including space utilization, occupancy management, HVAC controls, and energy management

- **Pattern rationale**:

  - Understand the behavior and control the assets and processes remotely, making them more predictable in the future

  - Provide a virtual representation of field devices

  - Avoid the need to have continuous data communication available between the central server and field devices

  - Provide a uniform interface for accessing information regarding a diverse set of information consumers

- **Related patterns**:

  - Rule engine

- **Assumptions**:

  - The usage of IoT sensor data and 3D visualization along with domain knowledge related to assets, processes and operations is required for effective DT implementation.

  - Devices are provisioned and relationships are defined between those devices. Additionally, the path/route for connecting to DT is preconfigured on the field devices when the device is installed or configured.

  - DT will have data-buffering mechanisms implemented and data will be normalized, filtered, and primed before being leveraged for data analysis/visualization purposes. Buffering is required both on the input side (data needs to be staged from multiple devices before being processed) as well as on the output side (data needs to be staged until the device can accept commands or connectivity is restored).

  - Data communicated by field devices is sent on standard protocols to DT. In the case of proprietary data formats/protocols, a field gateway is expected to perform the required protocol translations.

  - Data being transmitted follows regulatory/privacy norms, such as the GDPR.

  - The central server should have sufficient compute and storage capabilities to support the DT's functional needs.

- **Considerations**:

  - You should select a current and target DT maturity level for implementation and a roadmap for transitioning from the current to the target DT maturity level

  - The type of data selected and the rate of data capture will help simulate the device's behavior accurately, with related sensors helping to capture the required data

- **Anti-pattern scenarios**:

  - Scenarios where latency is considerable

  - Scenarios where data production, analysis, and consumption are to be done locally

Device management

The roles of the devices and how they are managed is the key differentiator between IoT and non-IoT deployments. The stages of device life cycle management include device provisioning (registration, activation, and commissioning) to de-provisioning. The notation used for device management in this book is shown in the following diagram:

Figure 2.6 – Notation for the device management pattern

Device management includes *firmware updates* for the field devices, either on an ad hoc basis (for example, patching security vulnerabilities) or in a planned manner (pushing the latest configuration and/or firmware with the updated feature set). Due to the constrained nature of IoT devices, special considerations are required while updating firmware:

- The device shouldn't be in the middle of a critical operation. Accordingly, the device's current state (busy, idle, and so on) should be shared with the central server at regular intervals.

- In some scenarios, the channel bandwidth is limited, so it is prudent to send a new version of the firmware in the form of packets or *chunks*. This will eliminate the need to send the complete firmware if there is a transmission error/data loss for a specific *chunk*.

Typically, the number of deployed field devices demonstrates a *hockey stick* style of growth once the pilot phase of the implementation of the IoT solution is complete. As a result, it is important to ensure that the *device management* module that's running on the central server is designed to be fully scalable. Additionally, to effectively manage a large number of field devices, it is prudent to segregate devices into *device families*. Segregation/categorization can be done based on physical location (mimicking the actual physical topology of the device, such as its country, county, village, and so on) or time (for example, the year the device was onboarded).

In addition to several devices, another challenge is that the diverse set of devices needs to be managed. Typical IoT deployments are done in a brownfield type of scenario, where new devices coexist with legacy devices. Legacy devices need to be managed and carefully planned as most of them won't support all the device management operations (such as firmware updates).

The device registration process typically follows the following sequence:

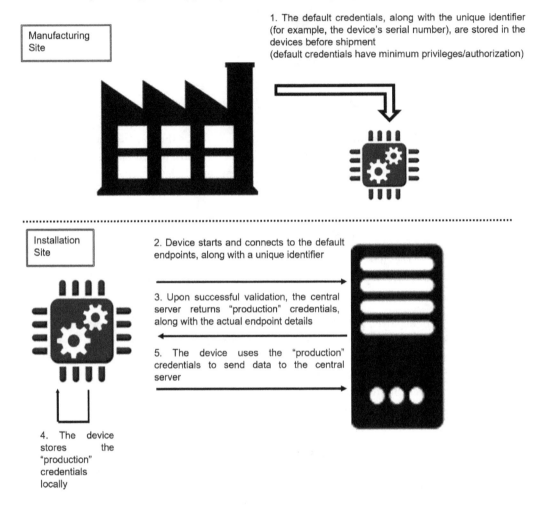

Manufacturing Site

1. The default credentials, along with the unique identifier (for example, the device's serial number), are stored in the devices before shipment
(default credentials have minimum privileges/authorization)

Installation Site

2. Device starts and connects to the default endpoints, along with a unique identifier

3. Upon successful validation, the central server returns "production" credentials, along with the actual endpoint details

5. The device uses the "production" credentials to send data to the central server

4. The device stores the "production" credentials locally

Figure 2.7 – Device provisioning flow

The following are the main functionalities that are expected from a device management solution:

- **Life cycle management**: The device management component is expected to manage the field devices, from initial provisioning to final decommissioning. Also, all communication between the devices and the central server should happen securely – only authenticated devices with proper authorization should be able to send data to the central server and receive commands. The timely rotation of security keys, certificates, and more, also needs to be ensured.

- **Configuration management**: The need to modify the behavior of deployed devices by pushing the relevant configurations from the backend is another key need as it helps make the required changes remotely and avoid unncessary and expensive truck rolls. Typical examples of configuration include changing the frequency of data reporting to the central server to optimize battery and/or channel bandwidth, requesting additional device state information for diagnostics purposes, and toggling features/functionalities on/off to conserve battery life.

  If changing the configuration parameter results in the device's state becoming unpredictable (for example, the device becomes unresponsive), the device should have built-in mechanisms to restart itself (using watchdog timers, for example) or roll back to the previous configurations.

- **Device state visibility**: Device management should present the state of the devices (the connectivity status, such as connected/not connected, the last connection or data transfer time, the battery status, and more), along with device metadata (for example, the date and time of device installation, physical location, current firmware version, and so on) in the form of a dashboard. In scenarios where very low device downtime is acceptable, device management should send timely alerts or notifications to the relevant team.

> **Important Note**
>
> Device management spans the complete life cycle of field devices (registration, activation, and commissioning) and also supports regular (feature releases) and ad hoc (security vulnerability patching) firmware releases.

Pattern summary

The patter summary for device management is as follows:

- **Problem solved**:

  - **Business**:

    - Streamline device onboarding, resulting in shorter deployment timelines

    - Bulk device onboarding to accelerate device rollouts

    - Third-party device onboarding will enable new business opportunities

    - Reduce the cost of field operations for devices via remote configuration/firmware updates

    - Optimize device diagnosis and troubleshooting with reduced cost of operations

    - Secure firmware/patch updates, leading to uninterrupted operations (avoiding security breaches)

    - Visualize the customer device topology and understand the physical and logical placement of one device in relation to others

- **Technical**:

  - Remote and secure **over-the-air (OTA)** services accessible to the remote operator

  - Better control and co-ordination due to reusable device management services

  - Built-in scalability to ensure future workloads are handled correctly

  - Abstraction of device-related nuances and capabilities

- **Usage context**:

  - Device onboarding and connectivity

  - Heterogeneity of devices and/or number of devices

  - Firmware update management

  - Applying device security scenarios

  - Troubleshooting devices with remote access

  - Manage device connectivity and inventory

- **Example/usage scenarios**:

  - DG management is a key requirement for the success of the IoT program, which involves multiple stakeholders, field operations, and software running on edge devices

  - To manage device life cycle operations and to differentiate them from device data management operations (data ingestion, data storage, and so on)

- **Pattern rationale**:

  - A centralized mechanism to ensure the life cycle management of field devices is required

- **Related patterns**:

  - DG

  - DT

- **Assumptions**:

  - None

- **Considerations**:

  - Device scale and diversity

  - Firmware upgrade frequency trade-off with available device battery life

- Remote access to devices (device tunneling to connect the device behind a firewall)

- Device topology and grouping to help manage devices at scale

- Firmware update of the DG, as well the end devices

- Event-based or periodic firmware updates

- **Anti-pattern scenarios**:

  - Infrequent updates of firmware and/or configuration

  - A very small number of field devices not justifying the overhead of device management

  - Scenarios where manual updates are feasible

Summary

This chapter introduced the core device-related patterns (DG, DT, and device management). These patterns will help you develop *end-to-end* IoT architectures (that is, scenarios where data is sent by the devices and commands are then sent back to devices so that the required action can be taken). At this point, you should be able to make decisions regarding what functionalities need to be implemented in the DG and what functionalities need to be implemented at the central server.

The next chapter will expand on this list of IoT patterns and include patterns that are implemented on the central server.

3

IoT Patterns for the Central Server

This chapter lists the architectural patterns that are deployed on a central server due to storage and/ or computational requirements and deployed on the edge (on-premises) or a cloud. These patterns provide insights based on the data generated by field devices, analyzing and enriching the existing data using additional data (from additional systems, such as enterprise systems).

These patterns help to extract insights, as well as automate certain actions (switching on the irrigation pump if the soil moisture goes beyond a defined threshold, as an example). Simple decisions (*if X, then do Y*) can be built into the business logic, but typically, any complex decision requires the evaluation of multiple input parameters and certain *prior experience*, which is where **artificial intelligence (AI)** and **machine learning (ML)** play a significant role. This chapter details how the data obtained from field devices is preprocessed to enable *data-driven* decision-making.

In this chapter, we will cover the following patterns:

- AI/ML integration
- The rule engine
- File upload
- Enterprise system integration

AI/ML integration

In IoT solutions, AI/ML technologies enable machines and field devices to simulate intelligent behavior and help make informed decisions with little to no human involvement. The notation that we will use in this book for AI/ML integration is as follows:

Figure 3.1 – The notation for an AI/ML integration pattern

AI/ML provides each IoT device with a *distinct personality* or *identity* that helps us to understand the overall context and allows it to act on behalf of the end user. In other words, another layer of abstraction is provided for both field-generated insights (data from sensors) and commands (to actuators), as shown in the following figure:

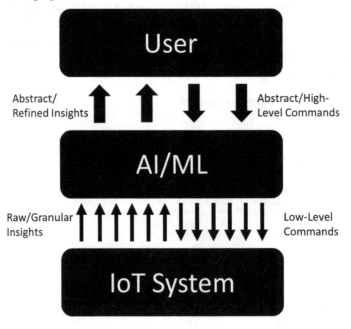

Figure 3.2 – AI/ML as an abstraction layer between the IoT system and end user

This pattern has multiple application scenarios, which are as follows:

- Cleaning dirty (as in incorrect, out-of-context) data and interpolating/extrapolating missing data.

- Making sense of the huge amount of data accumulated by sensors and eliminating false alarms. Considering the typical scale at which IoT operates, manual and separate monitoring of each device is not practical.

- Generating recommendations/actionable insights considering both real-time data streams and historical data.

- Measuring the calibration drift of sensors/actuators and automating associated rectifications.

- Determining the optimum place to perform analytics (or evaluate a rule/decision), whether at the edge or in the central server, balancing factors such as urgency, complexity, volume, latency, and battery/power status.

- Generating meaningful insights from accumulated data, instead of carrying out simple *data reporting*. For example, sourcing raw material from vendor *X* in place of vendor *Y* would help in improving the **Overall Equipment Effectiveness (OEE)** – detailed in *Chapter 7* – of a plant.

- Predicting performance bottlenecks and operational failures while eliminating/minimizing false alarms.

- Enabling use cases such as *object detection* at the edge and the monitoring and remediation of security threats in the central server.

- Automizing the detection of security threats, such as **Distributed Denial of Service (DDoS)**, and their remediation.

The preceding points can be handled by using simple rule-based algorithms (such as, *if event or data X, do Y*). However, the integration of AI/ML enhances both the breadth, which refers to the scope of analysis, and the depth of analysis, which refers to scenarios or events that were not considered during system design.

AI/ML model creation is typically done in a central server, owing to the high computational complexity involved, and then deployed at the edge or to a field device. Data received from the edge or this field device is continuously analyzed to further refine the model's accuracy. This *virtuous cycle* of model refinement can be better understood with the help of the following figure:

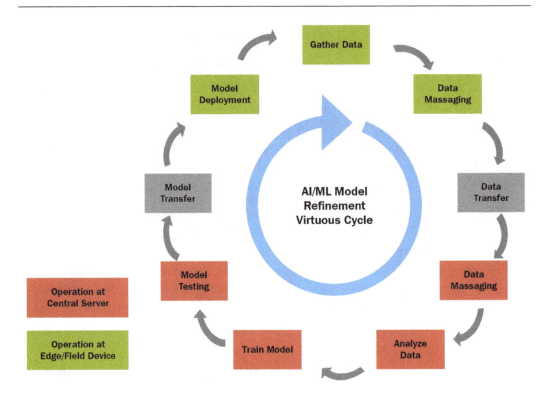

Figure 3.3 – Continuous refinement of the AI/ML model based on
the data/feedback received from field devices

Different types of AI/ML techniques relevant in an IoT context are summarized in the following diagram:

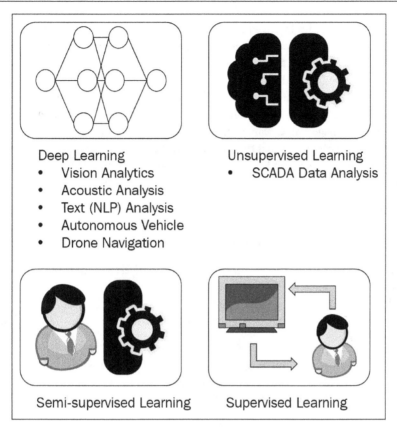

Figure 3.4 – AI/ML techniques relevant in an IoT context

The given variants of AI/ML all have different computing power requirements. Accordingly, not all the variants can run on IoT infrastructure. The following figure indicates how AI/ML variants are deployed in a typical IoT deployment:

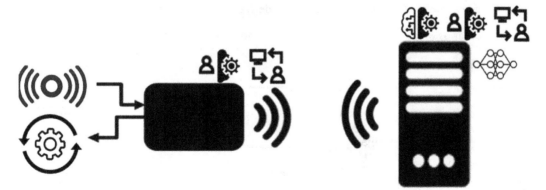

Figure 3.5 – A typical IoT deployment of different AI/ML variants

There are multiple AI/ML methodologies that can be used to fulfill different requirements. However, one family of ML *deep learning* techniques in particular (that of **zero-shot learning** and **few-shot learning**) is particularly well suited to IoT deployments. These deep learning techniques don't require a large dataset for training (or model creation) and rely on heuristics or metadata for arriving at a decision. *Zero-shot learning* indicates that this technique doesn't require any prior dataset (that is, no training data) for it to arrive at a decision (to identify a particular image, for example, there is no need to feed a set of images; an explanation of what an object of interest looks like is sufficient). Similarly, *few-shot learning* indicates that the technique requires minimal datasets (the dataset count typically ranges from one to five) and is primarily complemented by heuristics or resemblance data for it to come to a conclusion.

At first glance, zero-shot learning and few-shot learning may appear to be impractical given it is common to feed a large number of datasets to generate any practical ML model. However, zero-shot learning and few-shot learning mimic the way the human brain learns about concepts in the physical world.

Information such as *a dog is a four-legged animal whose skull and feet are smaller than a typical wolf – however, it has eyes larger than a wolf*, if fed to a child, can help them to identify a dog, assuming they have some understanding of what a wolf looks like. There is no need to give the child a large number of images of different dogs for them to recognize a dog. Zero-shot learning and few-shot learning work on a similar level.

Zero-shot learning and few-shot learning have relevance in IoT because field devices and device gateways are typically constrained from a computation and storage standpoint. As such, it is impractical to store and use a large number of datasets for training and subsequent model deployments.

Another example of few-shot learning would be the case where field devices are expected to recognize digits (0 to 9) within images. Here, traditional ML would involve feeding all the variants of the ways that digits can be typed or written to the model – samples can easily run into thousands, if not millions, which makes it impractical to execute the model on an edge or field device. However, making use of *relative semblance* can remove the need for requiring such a large dataset. In the case of digit identification, it is sufficient to feed heuristic or empirical information such as *digit 3 is roughly half of digit 8 and one-third of digit 5* for the edge or field devices to identify a digit.

A few IoT scenarios where zero-shot learning and few-shot learning can be used are as follows:

- Scene identification
- Edge analytics
- Object recognition
- Natural language processing
- Video analytics

Now let's take a look at the pattern summary.

Pattern summary

Let's take a look at the pattern summary for AI/ML integration:

- **Problem solved**:
  - **Business**:
    - Decision automation at the edge or in the central server
    - Use of data in planning and for increasing operational efficiency
    - Reduction and elimination of downtime
    - Enablement of mass customization and personalization
    - Creation of a person-independent recommendation system
    - Need for continuous refinement of the model
  - **Technical**:
    - Enablement of digital twins and digital threads
    - Service of data to the AI/ML engineers with the right data model
    - Validation of schema in the stream processing of IoT data
    - Application of the data model on the IoT edge devices
    - Model creation and deployment for constrained field devices or edge gateways
    - Need for actionable insights at the edge
- **Example usage scenarios**:
  - Capturing tags (data points) from the assets in the plant operations
  - Sending commands back to the devices based on the behavior of the assets
  - Serving analytics models for the data science teams
- **Pattern rationale**:
  - Creating or refining the AI/ML model in a central server (with relatively high computation and storage requirements) and deploying it to the edge (having constrained compute, power, and storage capabilities) for local decision-making and timely action
  - Periodic/event-based deployment of models at the edge, resulting in continuously maturing AI/ML models
- **Related patterns**:
  - Digital twin

- **Assumptions**:

  - Manual analysis of the IoT-generated data is not feasible due to the data's volume, velocity, veracity, and so on

  - Availability of data for model training and creation (except in the case of zero-shot learning and few-shot learning techniques, as described in earlier sections)

- **Considerations**:

  - A time series database for storing real-time telemetry information

  - An IoT broker to get the data transported from sensors to gateways to the central server

  - Usage of the in-memory cache to serve urgent data queries

  - Microservices-based architecture for supporting future scalability needs

  - Customizing models to compensate for computational and storage limitations of edge devices (typically, by reducing model accuracy and/or speed), as in zero-shot learning and few-shot learning techniques

  - Capability for scheduled model deployment (and rollback), especially in the case of a large number of field or edge devices

- **Anti-pattern scenarios**:

  - Rudimentary decision-making (*if this, then do that*)

  - Limited input or raw data

  - Batch or offline data analysis

The rule engine

The rule engine is essentially mapping between IoT events and actions that need to be associated with those events. In the IoT context, events are typically generated using sensors, and required actions are taken by an actuator. The notation used for this pattern in this book is shown in the following figure:

Figure 3.6 – The notation for a "rule engine" pattern

An interesting analogy is to compare the *rule engine* to the human mind, where the use of the five senses can be compared to the activation of sensors, and then corresponding actions are taken by body parts (hands or legs, for example), as shown in the following figure:

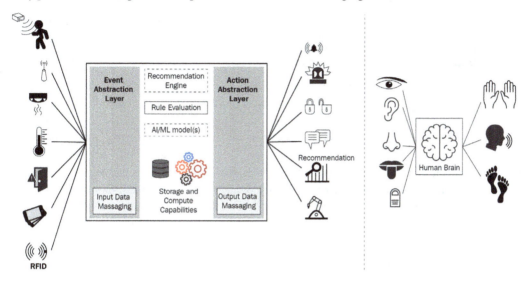

Figure 3.7 – A rule engine decoupling sensors and actuators and
its comparison with the human body/mind

Additionally, the *cause-and-effect* relationships in the IoT context are dynamic as well as complex, which can only be made possible by having an entity such as the rule engine in place. This is illustrated by the following examples:

- In an industrial context, the need may be for a buzzer to sound when an incorrectly assembled part is detected by installed video cameras (the events). However, eventually, the requirement may change to incorporate flashing lights on the assembly line and alerting the plant supervisor, in addition to (or instead of) simply sounding a buzzer.

- Currently, a home automation system is set to close the blinds and switch on the lights when darkness is detected outside the home. However, after some time, the user might want to only switch on the lights and leave the blinds as they are.

- In the healthcare space, a digital health platform might be supporting a weighing scale and glucometer only. However, with advancement in technologies, there might be a requirement to *plug in* additional devices, such as a defibrillator or an ECG. Having a rule engine would allow the easy configuration of rules as new devices are onboarded to the digital platform.

The information flow between sensors, actuators, and the rule engine typically consists of complex relationships. For example, the output from the rule engine can be a specific recommendation. Based on these recommendations, the rule engine may initiate further action either automatically or semi-automatically (depending on whether consent is required from a human party, as in *human-in-the-loop scenarios*). This is illustrated in the following figure:

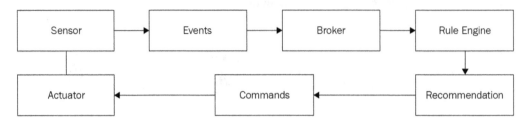

Figure 3.8 – The typical implementation of a rule engine will involve loops and complex interactions

From a deployment standpoint, the rule engine can be deployed at two levels, either **local rule engine** (**LRE**) deployment or **global rule engine** (**GRE**) deployment. LRE deployment (also referred to as *edge* deployment) helps in making quick and local decisions and is especially useful in scenarios where latency is a concern or connectivity is not completely reliable. LRE can also be seen as a *data feeder* to GRE.

GRE typically requires higher computational power, and hence, the ability to perform advanced analytics. Additionally, as multiple LREs feed into a single GRE, GREs can operate in a wider and more higher or global context compared to LREs. Typically, an LRE is hosted on the device itself or on a gateway. A GRE can be hosted on public or private data centers or the cloud. The relationships between GREs and LREs are illustrated in the following figure:

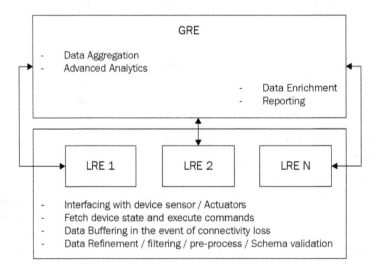

Figure 3.9 – Implementation of a rule engine will involve loops and complex interactions

The rule engine would need to be implemented at the *edge* or *locally* or on the *central server/globally* depending upon factors such as latency considerations, data volume requirements, and the current power/battery level of the edge devices. The configuration of these rules can be done via a *central server* or *global interface*, or it can be done via the *edge* or *local* interface (for example, if the edge or field device supports a **human-machine interface** (HMI for rule configuration).

At its core, a rule engine can be described as the following programmatic construct:

```
If ((sensor1 comparison1 threshold1) joining condition1 (sensor2
comparison2 threshold2) ... then
   Set Acutuator1 state as X
   ...


   End if
```

We can adapt the preceding generic construct to a more specific example. In a home automation context, it could consist of the following:

```
If((Thermostat1.CurrentValue > 20) and (Window1.CurrentState =
Closed)) Then
AirConditioning.CurrentState = ON
          AirConditioning.DesiredTemperature = 20
EndIf
```

As can be inferred from the preceding example, events can be correlated with actions as well as recommendations. Seen from one perspective, the rule engine helps to decouple sensors and actuators, very similar to the well-known *publisher/subscriber broker pattern*.

Providing recommendations rather than directly initiating actions is especially relevant in human-in-the-loop scenarios where it is desirable for a person to make a final decision as the impact of an action can be considerably higher – for example, where the safety and wellbeing of other people are involved, as in the healthcare domain. In the case of severe illness, it would be worthwhile for a rule engine to provide recommendations that can be reviewed by a medical specialist who has the final authority to make the call on the type of medication to be used and the required dosage. Recommendations can be provided using interfaces such as desktop or mobile interfaces or smart speakers (which are a form of actuators).

A rule engine can refer to historical data to arrive at a decision or it can be completely *stateless* – depending upon the use case scenario and, more importantly, on whether *prior experience* has any influence on the final decision. Any non-trivial decision-making process would require the integration of AI/ML models into the rule engine. Any decision that can't be taken based on simple *if X, then do Y* logic can be considered as a example of a non-trivial decision. AI/ML can be further refined where past decisions are monitored for any unintentional biases.

Rule engines of the future will act as trusted advisors who not only provide meaningful answers but also turn the question around so as to provide more engaging and meaningful conversation. Another related example could be a scenario where the rule engine results are curated to keep the interests of the user in mind. Essentially, the recommendations of the rule engine should not only be correct and accurate but also fit into an individual's belief system to be perceived as relevant and be more likely to be accepted. Hence, having a reasonably accurate estimation of *digital personality* plays a crucial role.

It is imperative that the final recommendations delivered can be adjusted for the *specificity level*; for example, in the case that a pointed answer is best suited to the context, recommendations should be filtered to provide minimal results (ideally, one). However, if the context in question demands a broad set (or range) of responses, then the response can be relatively more generic.

Related to the previous point, it can be noticed that the need for specificity may not be easy to deduce, as it involves objectively estimating the situational context (the culture and background, for example) of a user. Evidently, the situational context is inherently a complex thing to determine, as it depends on a diverse set of non-deterministic factors. One relatively easy (and highly effective) factor is taking instant feedback from the person for whom the recommendations are being formulated and feeding that input back into the rule engine, thereby continuously refining its effectiveness. This results in an interesting paradigm where the consumer of the rule engine output and the rule engine itself are continuously switching roles, in turn, lending credibility and maturity to the process of providing ever more personalized and relevant recommendations.

The availability of an optimized but relevant set of search results has a direct bearing on end user experience. Hence, rule engines will graduate from just being *text pattern-matching* machines to *smart, knowledgeable experts* that provide the most relevant recommendations.

To summarize, instead of spitting results out mindlessly, performing intelligent filtering (based on an individual's emotional state and overall situational context) would go a long way in devising next-generation rule engines. It would be desirable to intelligently massage and filter results before presenting them to users. Let's take a look at the pattern summary in the next section.

Pattern summary

The pattern summary for a rule engine is as follows:

- **Problem solved**:
  - **Business**:
    - Flexibility to add or update rules on a needed basis
    - Ability to configure rules for field devices on a central server and push it to field devices
    - Ability to define rules as mathematical and/or logical expressions for easy comprehension, therefore allowing non-technical users to make rule modifications
    - Admission of the rules to be used (with required modifications) across different domains and/or IoT use cases

- Ability to integrate more advanced technologies such as AI/ML for rule evaluation

  - **Technical**:

    - The presence of an abstraction layer for sensors and actuators minimizes the effort involved in the integration of new sensors and actuators, as code changes are localized in abstraction layers only.

- **Usage context**:

  - It is used where the behavior of events or actions individually or in relation to one other is expected to change.

  - A rule engine is primarily deployed in two ways:

    - **Fully automated**: The rule engine can make decisions and execute those decisions independently. This generally requires AI/ML integration.

    - **Semi-automated** (human-in-the-loop): Here, the responsibility of the rule engine is limited to analyzing input data and then providing recommendations to the user, who ultimately decides on the action to be taken. The generated recommendations can be sent in the form of push notifications on the user's mobile device for timely action.

  - The rule engine can be implemented either on the edge or in the central server, depending upon the complexity of the rules, as well as the compute and storage capabilities of field devices. Another consideration would be the case where data from field devices spread across vast geographical regions needs to be aggregated before a decision can be made, in which case, execution at the central server would also be required.

  - In scenarios where a specific rule needs to be selected from a set of available rules.

  - Typically, a rule engine would be implemented on a DG, which acts as a mediator between IoT sensors and actuators.

- **Example usage scenarios**:

  - Hosting or deploying at a central server and/or DG for configuring or executing rules in a structured manner

  - Decoupling IoT events and associated actions and commands, allowing extensibility for accommodating future events, actions, and commands

- **Pattern rationale**:

  - Provides a structured mechanism for decoupling events and actions on field devices and/ or at the central server, eliminating the need for custom plumbing logic that needs to be added to accommodate every new event and/or action, thus improving the maintainability of the system

- Helps to transform complex events into mathematical/logical expressions for ease of comprehension and modifiability

- Provides the ability to select a particular or optimum rule to be applied (from a set of potential rules) by integrating AI/ML technologies

- **Related patterns**:

  - None

- **Assumptions**:

  - Data is preprocessed and/or normalized before being fed to the rule engine (*input-side data massaging*). Similarly, data is converted into the form required by the actuators (*output-side data massaging*).

  - The rule engine would have the ability to maintain state information and wait for a state change in scenarios where a human is expected to provide final authorization on the recommended actions.

  - Time synchronization is required between the central server and field devices, especially in scenarios where evaluation defines time as one of the parameters.

- **Considerations**:

  - A rule would be deployed on a device gateway (field devices) or the central server based on the following factors:

    - The compute capabilities of field devices versus central server vis-à-vis compute requirements (compute or storage or both) for executing rules.

    - The time required for obtaining the rule engine results. It is important to understand that, in addition to the time required for rule execution at the central server, the delay introduced during data transfer needs to be considered, as per the following equation:

      Total = TDS (the data transfer time between the device gateway and central server)

      +

      TR (the actual rule evaluation time)

      +

      TSD (the data (rule results) transfer time starting from the central server and going back to the field device)

  - Depending on whether *rule selection* would be required in addition to *rule execution* or not, suitable AI/ML models need to be invoked at the device gateway or central server.

- The rule engine may integrate logging of the rules evaluated to provide an audit trail and for troubleshooting purposes.

- **Anti-pattern scenarios**:

  - Use cases where the list of sensors/actuators is relatively static

  - Use cases where relationships between events and actions are simple and/or static and can be hardcoded in the overall logic

File upload

The file upload pattern is relevant where real-time telemetry is not required or in cases where data needs to be fetched from legacy and non-connected devices. This pattern is also used to push the firmware to the field devices. The notation used for this pattern is shown in the following figure:

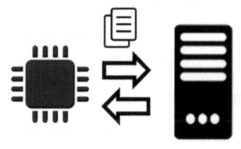

Figure 3.10 – The notation for a file upload pattern

Some scenarios where this pattern can be used are as follows:

- Video stream processing is to be done in a central server.

- Legacy systems in industrial, healthcare, or energy domains output data in a physical file and it is not possible to parse or interpret the generated file on the device due to computation and/or storage constraints.

- Certificates and keys are required for secure connection to the central server.

- Scheduled or ad hoc firmware update of devices.

File upload is generally done via packetizing the file content to optimize bandwidth usage. A typical algorithm for this is shown in the following diagram:

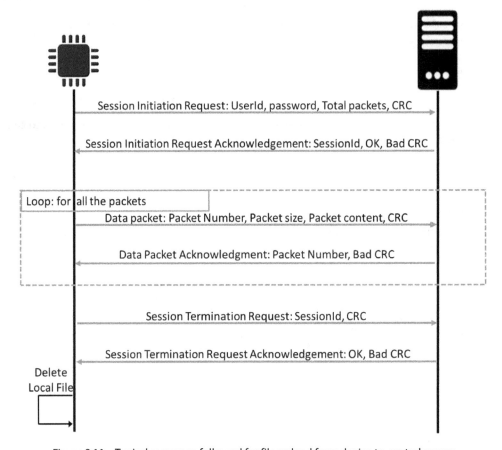

Figure 3.11 – Typical sequence followed for file upload from device to central server

Although the preceding diagram shows file upload from the device to the central server, a similar sequence would be required for the central server to send the file to the device. Also, for simplicity's sake, the scenario of a bad **cyclic redundancy check** (**CRC**) error is not depicted. In such cases, the device would attempt to resend the packet until a successful acknowledgment was returned by the central server. Let's look at the detailed pattern summary next.

Pattern summary

Let's look at the pattern summary for file upload:

- **Problem solved:**

  - **Business:**

    - Scenarios in which sending regular/real-time telemetry data is not required

- Scheduled (addition of new features and capabilities) or ad hoc (to fix a critical bug fix or to plug a security vulnerability) release of firmware

- Audio or video feeds such as surveillance of factory activities, sports feeds, less frequently compressed/ZIP files such as patient treatment files, or drone data ingestion

- Large payload data, such as smart city scenarios for traffic data

- Controlling bandwidth usage by controlling the packet size

- Accumulating and sending data during off-peak hours

- **Technical**:

  - Overcoming the payload size limitations for file transfers

  - Rotating compromised or obsolete certificates of field devices for ensuring secure communication with central server at all times

  - Delivery assurance with checksum at the packet level

  - Using less chatty protocols (FTP over HTTP) to optimize bandwidth usage

- **Usage context**:

  - Scenarios where real-time telemetry data collection is not a requirement.

  - Scenarios where network bandwidth is at a premium and chattiness of the telemetry protocol (HTTP, for example) is not acceptable.

  - High-speed data is captured in binary mode and is decoded and processed at the central server.

  - Real-time data processing is not a requirement.

  - Scenarios where large payload data is required to be sent in packets or batches.

- **Example usage scenarios**:

  - Surveillance camera feeds in factories, buildings, or retail shops

  - Audio acoustics analysis for leak detection in utility pipes

  - Patient treatment files and machine configuration upload in the healthcare domain

  - Autonomous vehicles where sensor fusion can result in a large binary unstructured payload that needs to be uploaded

  - Certificate rotation in case of mutual authentication

- **Pattern rationale**:

  - Less frequent but large payload size

- Optimum usage of network bandwidth

- Offline/batch processing

- Decoupling of data ingestion and processing

- Decoupling of data collation and transfer

- **Related patterns**:

  - Digital twin

- **Assumptions**:

  - Existence of data storage space to store an intermediate file(s).

  - File data format agreement between the device and central server.

  - A file transfer trigger, either periodic or event-based (when a predetermined file size is reached, for example), has been configured.

- **Considerations**:

  - Data processing can be done at the edge or on the field device without requiring a file transfer.

- **Anti-pattern scenarios**:

  - Large binary payloads (video feeds, for example) can be processed at the edge and only events need to be sent to the central server.

  - Real-time data processing and event generation.

Enterprise system integration

To derive maximum value from IoT implementation, the IoT system needs to integrate with existing software systems in an enterprise. In fact, integration with external systems allows for the interaction between the cyber and physical worlds – IoT systems representing the physical world and external systems residing in the cyber/virtual world. The notation used for this pattern is shown in the following figure:

Figure 3.12 – Notation for an external system integration pattern

As is the case with any general integration, integration of an IoT system with another enterprise or legacy system results in benefits that far exceed what we would have if these systems were operating in isolation (the whole being greater than the sum of its parts). Integration can enable automation of *end-to-end* workflows, remove data duplicity, enhance decision-making quality, or eliminate the possibility of relying on stale data.

Data synchronization between IoT and other enterprise systems may be enabled by either a *data push* or a *data pull* or both. The nature and level of this integration will vary from domain to domain and from one use case to another. For example, *home automation* use cases tend to have less need for integration with other systems compared to industrial use cases. As IoT implementation enables automation and hence tends to replace or augment existing workflows, processes, and the like, it is only natural that the IoT data needs to *plug into* existing workflows and processes. Since existing workflows are built on top of existing or legacy systems, the need for *enterprise integration* becomes urgent.

The frequency at which integration logic is invoked will depend on multiple factors, such as use case requirements and overheads introduced by the *data synchronization* operation. Depending on the requirements, synchronization can be done in synchronous mode (real-time sync), or it can be enabled in batch mode, where an operation runs at a scheduled frequency and time. Generally, such integrations are enabled without any direct involvement of the end user.

IoT systems are typically integrated with the following types of external systems:

- **Customer relationship management (CRM)** systems
- **Supply chain management (SCM)** systems
- Business intelligence and analytics
- Human resources data
- Systems storing device or user metadata
- **Enterprise resource planning (ERP)** systems
- Systems providing auxiliary information (such as weather information or satellite imagery), which can augment or refine insights generated by IoT data

Integration requirements are realized by invoking the APIs exposed by enterprise systems. In a scenario where such integration APIs are not available, a wrapper on the legacy system is created, which can then encapsulate the nuances of the legacy systems.

External systems can complement both local and global rule engines, as shown in the following figure:

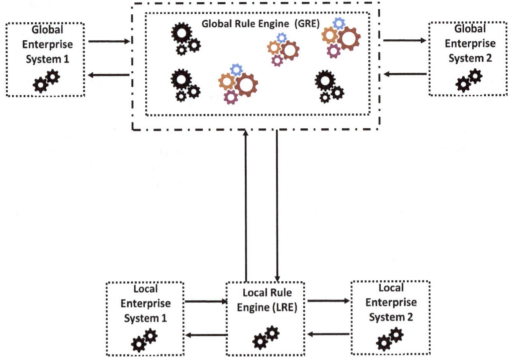

Figure 3.13 – The need to integrate with external systems can
exist at both a local and global rule engine level

An IoT system can integrate with other external systems in a variety of ways (depending upon the application and use case needs).

Although the following figure shows integration with respect to the global rule engine, similar integrations are possible at the local rule engine level as well:

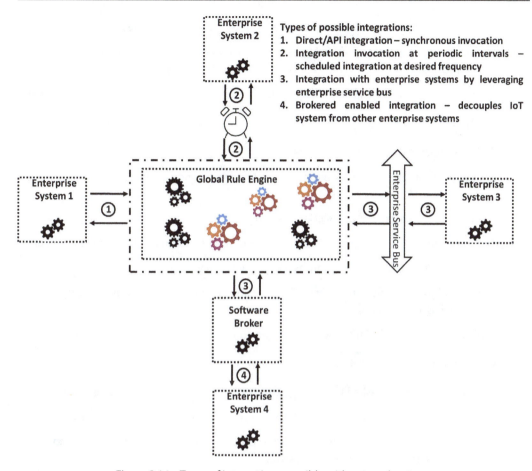

Figure 3.14 – Types of integrations possible with external systems

Now let's look at the pattern summary.

Pattern summary

The pattern summary for enterprise system integration is as follows:

- **Problem solved**:

  - **Business**:

    - Integrating IoT data with existing enterprise applications, with benefits as follows:

      a) Better insight and decision-making abilities

      b) More accurate business insights

      c) Cost or effort reduction due to reduction of data reconciliation efforts

- Conceptualizing and developing richer and more innovative use cases

- Accelerating an enterprise's digital transformation journey by automating existing workflows and processes vis-à-vis force-fitting entirely new tools, processes, and workflows

- Enabling data cleaning and data massaging by using device metadata (from enterprise systems)

- **Technical**:

  - Synchronizing data between systems

  - Choice of multiple integration options, such as synchronous, scheduled, broker-based, and so on

  - Enabling both data- and application-level integration

  - Scheduling integration for handling compute-intensive workloads during off-peak hours

  - Flexibility to implement real-time as well as batch integrations

  - Decoupling the evolution of core IoT system from downstream external systems using broker-based integration

  - Flagging and remediating data gaps between different systems as soon as possible

- **Usage context**:

  - IoT data needs to be pushed to existing external, enterprise, or legacy systems.

  - IoT data needs to be enriched with metadata by pulling data from external systems.

  - The transition of the external system's workflow from stage to stage is based on the information from sensors as reported by the IoT system.

  - Offline data integration (as in data exported from one system, followed by importing it into another system using a two-step process) is not practical or suitable.

- **Example usage scenarios**:

  - **Worker safety use case**: An employee's personal data is pulled from an employee database (for example, an HRM external system) to notify the supervisor if a worker had a fall (as detected by a fall sensor).

  - **Smart manufacturing use case**: The count of available parts falls below a defined threshold; an order for fresh supply needs to be initiated by getting vendor information from a **vendor management system (VMS)**.

- **Pattern rationale**:

  - IoT data has limited value unless it is enriched or integrated with other sources of enterprise data.

  - An IoT system senses an environment (and/or the condition of an entity within that environment) and generates events, which are fed into external systems to trigger workflow stage transitions (for example, a **Goods Received Note (GRN)** is initiated in the case that a shipment is detected at the entry gate).

- **Related patterns**:

  - Rule engine (to trigger enterprise workflows)

  - AI/ML integration, to predict events and trends (for example, supply and demand trends) and to trigger proactive as opposed to reactive actions

- **Assumptions**:

  - The system hosting the rule engine is capable of handling the additional load for supporting integration requirements.

  - External systems expose interfaces (APIs, for example) for data integration needs.

- **Considerations**:

  - Needing to select the optimum integration type based on current and future needs

  - Selecting the ideal synchronization frequency based on factors such as data concurrency, end user expectations, and the off-peak load window

  - Minimizing the impact on the *as is* operation of external systems

- **Anti-pattern scenarios**:

  - Scenarios where organizations can tolerate siloed data collection/analysis

Summary

This chapter introduced architectural patterns (AI/ML integration, the rule engine, file upload, and enterprise system integration) that are typically deployed on a central server. The patterns in this chapter, as well as those detailed in previous chapters, will empower you to architect any IoT application. Subsequent chapters will show us examples of how these patterns are combined to solve complex problems in different domains, starting with the next chapter, where we will discuss two specific use cases in the consumer domain – home automation and a smart egg boiler – and understand how the architectural patterns we've previously learned about can be applied to develop interesting use cases.

Part 2:
IoT Patterns in Action

The patterns detailed in the previous part can be mixed and matched to realize IoT use cases. This part provides proof of efficacy for these patterns to satisfy unique needs and implement use cases in diverse domains, including consumer goods and home automation, retail, transportation, manufacturing, and agriculture.

This part comprises the following chapters:

- *Chapter 4, Pattern Implementation in the Consumer Domain*
- *Chapter 5, Pattern Implementation in the Smart City Domain*
- *Chapter 6, Pattern Implementation in the Retail Domain*
- *Chapter 7, Pattern Implementation in the Manufacturing Domain*
- *Chapter 8, Pattern Implementation in the Agriculture Domain*

4

Pattern Implementation in the Consumer Domain

We learned important architectural patterns in previous chapters; this chapter introduces use cases for these patterns that are relevant to the consumer domain. Although numerous use cases exist in the consumer space (e-health, elderly care, pet tracking, energy management, safety and security, robot vacuum cleaners, etc.), the current chapter will detail just two use cases – **home automation** and a **smart egg boiler** – to give perspective on the nuances that are involved while implementing consumer IoT use cases. Home automation is an existing use case, whereas the smart egg boiler is an innovative use case – giving credence to the fact that known use cases can not only be implemented by leveraging IoT patterns but can also be used to implement hitherto unknown solutions.

Use case – deploying home automation

Typical home automation deployments provide a combination of the following feature sets:

- Controlling and monitoring appliances, such as air conditioners, refrigerators, or heaters
- The controlling of lights (*on/off*, a change of color or brightness, or *on/off* based on a person's presence)
- An action based on a detected event (for example, an alarm/buzzer sounding when smoke is detected)
- The remote operation of doors and windows
- The detection of an intruder that triggers an alarm
- Conversations with smart speakers

The following figure gives a representational view of home automation implementation:

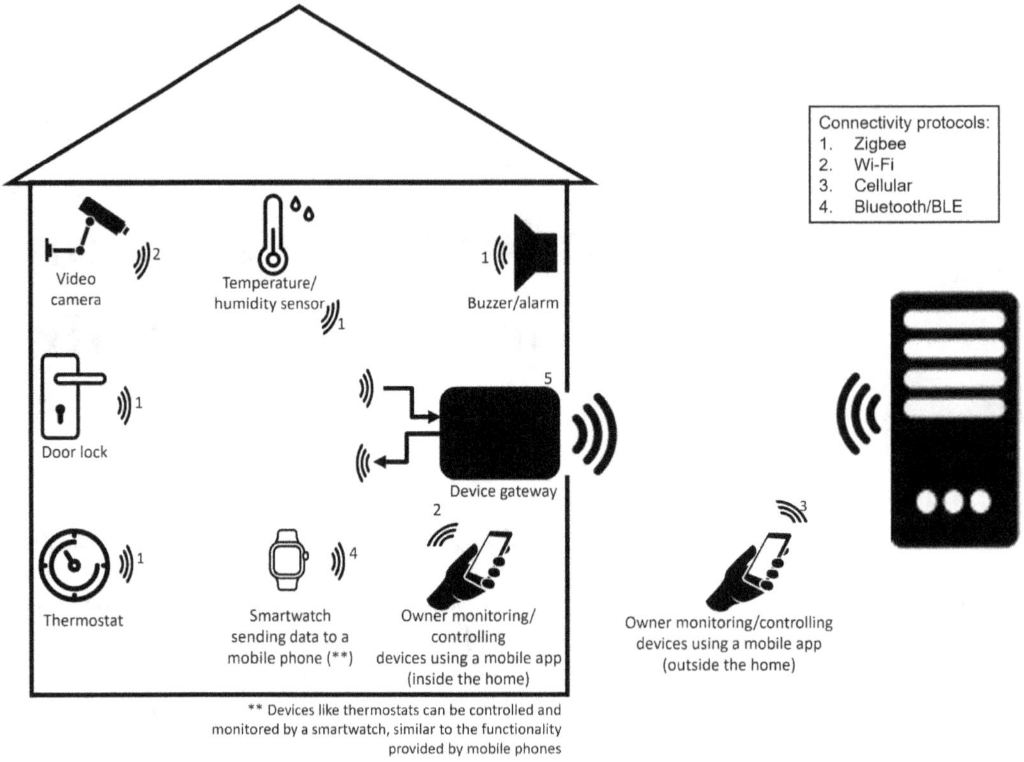

Figure 4.1 – Sensors/actuators in a home automation use case

As you can see, devices that are battery-operated and have low bandwidth requirements typically transfer data on an energy-efficient protocol (for example, Zigbee), whereas devices such as a video camera send the data to a **Device Gateway** (**DG**) over protocols such as Wi-Fi.

Figure 4.1 also shows a mobile phone as a primary interface to control and monitor *smart home* devices; new interfaces (for example, smart speakers) are also emerging on the market – for example, the latest trend is the use of a smartwatch to control or monitor smart devices and appliances in addition to the mobile phone. However, due to the smartwatch's smaller interface, only a subset of mobile app functionality (e.g., critical alarms or notification events and turning main appliances on and off) can be made available on a smartwatch.

In general, the adoption of home automation use cases has been at a slower pace compared to other IoT use cases (for example, data-driven manufacturing). Some of the reasons include installation complexity (especially for non-tech-savvy users), limited interoperability between smart home devices from different manufacturers (although there are initiatives such as Matter from the **Connectivity**

Standards Alliance (**CSA**) that are expected to bring a desired standardization to various smart home devices), and the perception of home automation as *good to have* rather than a *must-have* requirement.

The benefits of installing smart home devices include the following:

- **Energy savings**:

  - You can monitor the consumption at a granular (appliance) level and replace/repair a power-hogging appliance.

  - Turning off appliances after leaving home.

  - In addition to raising alerts if energy consumption goes beyond a threshold, you can provide recommendations regarding how energy consumption can be optimized by leveraging AI/ML capabilities (for example, scheduling the operation of energy-hogging appliances during off-peak hours to benefit from demand-based pricing plans, or alerting a user if the temperature is not changing at a desired rate after switching on the air conditioner, which possibly indicates an open window or door).

- **Convenience**:

  - The opening of a garage door from a distance

  - Automatic control of ambient temperature/light

  - Storing and adhering to the user's preference with regard to temperature, light, and other ambient conditions

  - Scheduling routine tasks (for example, starting and stopping garden sprinklers)

- **Security**:

  - The 24/7 monitoring of kids and pets.

  - The user receives a notification on their mobile phone regarding possible intrusion attempts.

A detailed description of the use case

It can be seen from *Figure 4.2* that monitoring and control are possible from both inside as well as outside the home; however (in some cases), users don't want data to be transmitted outside their home boundaries, due to privacy and security concerns (a *smart home* can be maliciously used to identify and exploit the user's absence and gain unauthorized access). In these cases, DG acts at a local level only, and there is no connectivity with a central server. In order to cater to safety and security as well as the privacy concerns of users, this use case can be implemented in two deployment formats:

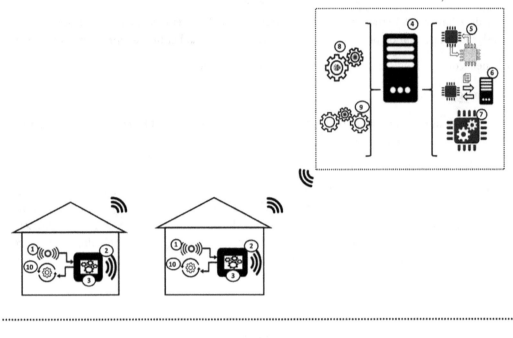

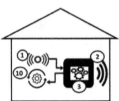

Figure 4.2 – The realization of a home automation use case by leveraging IoT patterns

These two types of deployment are as follows:

- **Deployment without a central server (standalone deployment):** This deployment is relevant for users who have security/privacy concerns and are not comfortable sending data outside a local/home network, although this deployment can provide only a subset of possible home automation features/functionalities. In this case, it is not possible to obtain refined AI/ML models in a continuous fashion from a central server; hence, the models deployed are static in nature (or a model is only updated when a device is shipped for repair to the manufacturer). This deployment model is shown in the bottom portion of the preceding figure.

- **Deployment with a central server:** This refers to the implementation where a DG installed in different homes sends data to a central server for collation, aggregation, analytics, and so on. This will support additional features (over and above what is possible with only standalone deployment) – for example, a recommendation engine (leveraging AI/ML technologies) that suggests ways and means to reduce energy consumption, automatic DG firmware updates, and comparing energy consumption with fellow users that have similar appliance usage.

As a user's private data is transferred over public networks, end-to-end security becomes a paramount requirement. Security considerations are also critical here, as a malicious actor might tap into a communication channel to force actions such as unlocking a door.

The main components of home automation as shown in *Figure 4.2* are as follows:

1. **Sensor:** Typical sensors used for home automation include fire/smoke/CO_2, water leak/moisture, water overflow, and window/door open/closed detectors, a video camera (internal or at the main entrance, integrated in a doorbell), motion, light, air quality, energy meter, and sound. These sensors generally communicate their status to the DG over protocols such as Wi-Fi, Zigbee, Z-Wave, and BLE/Bluetooth.

2. **Device Gateway:** In home automation deployment scenarios, a DG may or may not be connected to a central server, as shown in *Figure 4.2*. In the case of local deployment (without central server connectivity), the DG generally acts as a bridge between the sensors and actuators and executes a set of rules (for example, in case of a fire, sounding an alarm). It would also expose the internal state as well as receive commands via a set of predefined APIs that would be leveraged by mobile applications.

   In the case of connectivity with a central server, additional features can be enabled – for example, a user can compare energy usage with other residents in the locality, the DG would be updated with the latest firmware, and AI/ML components on the central server would help predict energy usage and provide recommendations about how to conserve energy and reduce energy-related charges.

3. **Local Rule Engine (LRE):** If DG functionality is consumed via a browser, content related to web pages (for example, HTML and CSS) is also hosted in the local DG only. Additionally, there is a provision to configure and store a set of instructions that would be invoked by a user on a regular basis – for example, instead of starting air conditioning at a desired temperature and closing the windows, both commands can be stored as one *mood* or *scene* and invoked as a *combo* command, rather than sending separate commands. Since there is no connectivity with the external world, typically the firmware in DG is updated via alternative means (for example, using a USB).

4. **Central server:** A central server will aggregate data from multiple smart homes and host components such as the digital twin, GRE, AI/ML model creation, and deployment logic. The central server will also host functionality to deploy firmware upgrades/patches onto the gateways installed in individual homes and manage security requirements – for example, certificate rotation.

5. **Digital Twin (DT):** A DT will act as a virtual representation of every home. The user can view the current state of their home sensors and is able to initiate required actions (for example, turning off the air conditioning if it was inadvertently kept on). To preserve privacy and ensure

security and safety, DT access will be enabled only for authorized end users, and **anonymization techniques** will be leveraged to ensure that data is not used maliciously.

6. **File upload**: In this use case, a file upload pattern would be used to enable firmware upgrade to a DG and (in some cases, such as downloading security patches or providing new features and functionality) to end devices (sensors and actuators) and to enable certificate rotation.

7. **Device management**: This pattern is used to perform the following functions:

   - Onboard and deboard DGs and associated end devices

   - Monitor the connectivity status of individual DGs

   - Fetch the system state and issue troubleshooting commands

8. **AI/ML integration**: AI/ML models created on a central server and deployed on the DG will help to analyze the usage habits of occupants, predict future usage trends, provide recommendations with regard to appliance usage, and automate control of devices such as a thermostat to reduce energy bills.

   This component would also automate the setting of home appliances (brightness levels, temperature settings, and so on) as per the user's past preferences. For richer use cases, this component would interface with an **external system integration** pattern – for example, knowing local weather conditions by fetching data from a weather service will help to determine the ideal duration/time for which home garden sprinklers need to be turned on (i.e., there is no need to water the garden in the morning if rainfall is expected in the afternoon). Similarly, room heating can be reduced if a sudden warm climate is expected.

   As a user's physical activity, as well as sleep quality/duration, is tracked via a smartwatch, AI/ML integration can help nudge them to adopt a more active/healthy lifestyle.

9. **External system integration**: A central server can be integrated with external systems such as a weather information service and an energy utility company billing/charging system to realize richer end-to-end use cases. Similarly, integration is possible with external **blockchain** systems (the usage of blockchain in an IoT context is discussed in greater detail in *Chapter 12*) to ensure secure billing transactions.

   Integration with a weather information service would be required to know future weather conditions to optimize when a garden is watered, as mentioned earlier. Integration with an energy utility company's system would help to understand charging patterns and schedule the operation of non-critical equipment (for example, a washing machine) during off-peak/low charge periods.

10. **Actuator**: An actuator will help to execute (actuate) commands such as turning on a coffee pot, setting a target temperature for air conditioning equipment, and locking or unlocking a door. Generally, these commands are issued locally by an LRE. In very specific cases, the actuator state can be set by a central server (for example, when scheduling the operation of a washing machine), after taking due authorization from the user.

This section detailed a home automation use case and how it can be implemented using patterns learned in previous chapters. In the next section, we will explore how a smart, innovative, and connected product can be developed using IoT patterns.

Use case – a smart egg boiler

There are numerous egg boilers on the market; however, all of them suffer from one basic limitation – they treat all eggs in a batch the same (using the same quantity of water, boiling duration, and so on). However, even eggs from the same hen can have different characteristics. But for the best taste and texture of a boiled egg, each one needs to be treated differently, and boiling conditions need to be customized as per the unique internal and external characteristics of the egg:

- **External characteristics** include the size, shape, shell color (indicating the hen pedigree), and altitude at which the egg is boiled.

- **Internal characteristics** include the texture/density/viscosity of the yolk and albumen, egg age (the time difference between the hatching of the egg and the time of the egg being boiled), shell thickness, and proportion of egg white versus egg yolk.

This use case suggests an innovative boiling process for each egg by positioning it in its individual compartment, thereby providing a unique boiling environment/condition(s). We customize the boiling operation of each egg by having an isolated and insulated compartment and by subjecting each one to its unique boiling conditions. This is achieved by first scanning each egg for its unique characteristics – internal as well as external.

The solution mentioned in this section will measure internal characteristics by inserting a minuscule camera within the egg body (refer to *Figure 4.3*). This camera will serve two purposes:

- Help to gauge ideal boiling parameters (boiling time and water quantity required) for each cell by analyzing the internal characteristics (for example, the proportion of egg white versus egg yolk).

- Provide a real-time view (video feed) of egg internals during the boiling operation. This will help a user to monitor the egg's internal state and (if required) pre-emptively cancel the boiling operation. Similarly, it will enable the *smart boiler* circuitry (by leveraging video analytics) to perform real-time condition monitoring and stop the boiling operation once the desired/ideal state is reached.

Although there might be some similarities between egg boilers available in the market, the smart egg boiler detailed in this section has the following unique characteristics:

- The proposed solution is a completely independent/standalone operational unit.

- The proposed solution customizes the boiling operation with the specific boiling requirements of each egg, considering the unique characteristics of each individual egg.

Important note

The smart egg boiler is a futuristic use case that demonstrates the *art of the possible* by using patterns listed in the previous chapters. However, this use case might not be immediately viable from a business or commercial standpoint due to cost, maintenance requirements, market acceptability, and other similar reasons.

Now that we've discussed the basics, let's look at this use case in further detail.

A detailed description of the use case

The key components of this innovative use case are illustrated in the following figure:

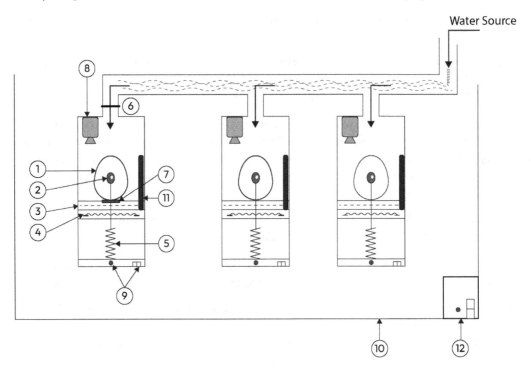

Figure 4.3 – An operational diagram of the smart egg boiler use case

The details of the components of the operational diagram are as follows:

1. An egg placed in an individual compartment.
2. A minuscule video camera inserted into the egg. A pH sensor to determine the egg age is also attached to the minuscule camera.

3. Water present in the individual compartment to generate steam. The quantity of water for a particular boiling session is adjusted by checking the current water level, by using a water level sensor (component 11) and sending instructions to the controlling valve (component 6).

4. The heating element used to boil the water and generate steam.

5. The mechanical spring that is used to prick the eggshell using a pin and then insert a video camera (refer to *Figure 4.5*).

6. The lever to control the flow of water/steam.

7. The weight sensor to measure the weight of the egg. The weight and dimensions as determined by a camera (component 8) are used to determine the egg density. Weight, dimensions, and density are used to gauge the egg's ideal boiling conditions. This requires past historical data as well as deploying an ML model onto the boiler's firmware.

8. A video camera is used to determine the external characteristics of the egg (the shape, size, and pigmentation). The input from this camera as well as the minuscule camera (component 2) are used to accurately determine the internal and external characteristics of the egg. This video camera constitutes two cameras (one 2D camera to determine the shape and pigmentation and another 3D camera to determine the egg size).

9. These are the switches and the **Light Emitting Diodes (LEDs)** provided to indicate the status of the boiling operation for the individual compartment.

10. The outer casing of the smart boiler within which individual compartments are placed. The number of individual compartments (in *Figure 4.3*) will vary, based on the number of eggs to be boiled in one batch.

11. This is a **sensor hub** that has a **water level sensor** to determine the current water level and remove the possibility of underflow or overflow, a **temperature sensor** to determine the current temperature of the compartment, and a **pressure sensor** to determine the current pressure conditions.

12. This is the instrument panel located on the main unit to indicate the operation of the overall unit and provide other controls, such as the *ON/OFF* switch.

The water level in each of the compartments is controlled by a combination of a water level sensor (component 11) and the water flow regulator (component 6). The boiling of water is performed by a heating element (component 4). The complete assembly is contained in the egg boiler body (component 10).

A high-level context diagram of the solution is shown in the following figure:

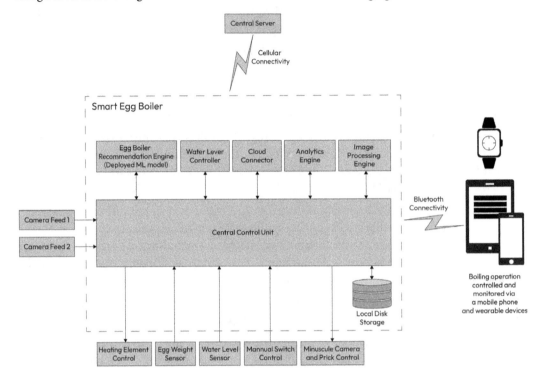

Figure 4.4 – A high-level diagram of a smart egg boiler

The following diagram details the step-by-step process regarding how an egg is pricked:

Step 1: The pricking tool and minuscule camera are placed on a plate that can move horizontally as well as vertically

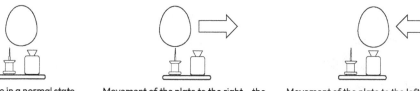

The plate in a normal state with the egg in a raw state

Movement of the plate to the right – the pricking tool is directly below the eggshell

Movement of the plate to the left – the camera is directly below the eggshell

Step 2: The pricking tool is used to prick the egg by moving the plate vertically in an upward direction, thereby making a vent for the minuscule camera to be inserted into the eggshell. Once the prick is made, the plate returns to its normal state

The plate in its pricking state

The plate in its normal state

Step 3: The minuscule camera is then inserted into the eggshell via the vent made previously. Depending on the degree of the plate's vertical movement, the camera can be adjusted to view any part of the egg

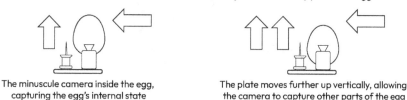

The minuscule camera inside the egg, capturing the egg's internal state

The plate moves further up vertically, allowing the camera to capture other parts of the egg

Step 4: Once the desired boiling state is reached (as determined by the locally hosted machine learning model), the plate again moves to its normal state, and the boiling cycle is complete

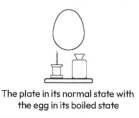

The plate in its normal state with the egg in its boiled state

Figure 4.5 – The operation of the pricking tool and the minuscule camera

Ideal boiling conditions are initially configured by the egg boiler manufacturer. However, as video analytics forms a key part of the solution, these boiling conditions are continuously refined, based on the actual usage by end users (using mobile devices). The end users rate the output quality of each boiled egg, and this feedback is used to further refine the boiling operations. The training of these ML models is done on a central server and then deployed to the egg boiler device.

This smart egg boiler serves multiple benefits over traditional egg boilers:

- It gives a more holistic view of the egg-boiling process.

- It prevents eggs from breaking and spoiling the compartment.

- It provides an ideal boiling operation as per the egg's characteristics, resulting in better-quality boiled eggs.

- It optimizes water as well as energy consumption – you boil each egg as per its unique requirements, rather than subjecting all eggs to a similar boiling cycle (*condition-based* boiling rather than *time-based* boiling).

Realizing the use case

The implementation of the smart egg boiler is detailed in the following figure:

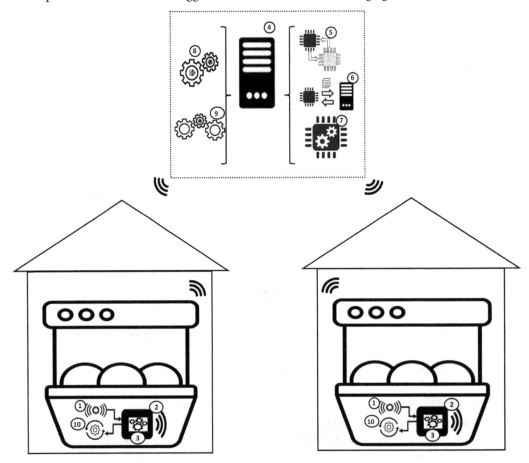

Figure 4.6 – The realization of the smart egg boiler use case by leveraging IoT patterns

The main components of the smart egg boiler solution (as depicted by the numbers in the preceding figure) are elaborated here:

1. **Sensor**: The unique requirements of the current use case can be met by using a diverse set of sensors:

    - A **water-level sensor** will determine the current water level in the boiling compartment and avoid water overflow or underflow. There is a variety of water level sensors available on the market that communicate using diverse communication technologies (Wi-Fi, Zigbee, analog, **Inter-Integrated Circuit** (**I2C**), and so on). However, in the current context, a water-level sensor communicating with the DG over a wired channel (analog or I2C) is the most appropriate.

    - A **minuscule 2D camera** is used to determine the shell thickness, the internal characteristics (for example, yolk texture), and whether the desired state of the egg has been reached or not. This camera is inserted via a spring action (refer to *Figure 4.5*). An additional function supported by this camera is to provide a real-time video feed to the user so that they can see the boiling state in real time and (if required) preemptively cancel the ongoing boiling operation. This camera sends the video feed to the DG over a wired (Ethernet) or wireless (Wi-Fi) channel.

    - One **additional 2D camera** is required to determine the external characteristics of the egg (the shape and pigmentation). To determine the size, a **3D camera** is required. Both cameras will send the feed to the DG over a wired (Ethernet) or wireless (Wi-Fi) channel.

    - A **pH sensor** is required to estimate the egg's age and is attached to the minuscule camera. Normally, the output of this sensor is fed into the DG using a wired channel, and the data can be sent in analog format or by using a digital interface (for example, the I2C protocol).

    - A **weight sensor** will help to determine the egg's weight, which will be used to estimate the egg's density. Like the water level sensor, the weight sensor is available in a rich set of connectivity options (**Bluetooth Low Energy** (**BLE**), Wi-Fi, Zigbee, analog, I2C, and so on); however, to optimize cost and power usage, the BLE protocol is suitable here.

    - The presence of a **temperature and pressure sensor** ensures that the egg boiler maintains temperatures within a defined threshold and reduces the possibility of egg breakage during boiling operation. Typically, both these sensors are housed in a single sensor assembly. Temperature and pressure sensors are supported on multiple communication protocols such as BLE, Wi-Fi, Zigbee, analog, and I2C. In the current use case, a wired connection to the DG with the I2C communication protocol is the most suitable.

2. **Device Gateway**: The DG here handles multiple functionalities such as the following:

    - **Receiving and analyzing data** from attached sensors. As the sensors interact with the DG over multiple protocols, they should support all those protocols.

- **Enabling local and remote connectivity** Local connectivity is required to communicate with locally paired devices (mobile phones, tablets, a smartwatch, and so on). This pairing, using the BLE communication protocol, helps users to monitor the operation of the smart boiler, set preferences or settings (soft/medium/hard-boiled, setting the water level, enabling a child lock, and so on), and configure the settings.

- Additionally, the DG is equipped with Wi-Fi or a cellular module to **enable connectivity with a central server**. Central server connectivity is required to receive firmware updates, download the latest ML models to ensure the best boiling conditions, send feedback from the user regarding boiling results, and so on.

- **Sending control instructions to actuators** to regulate the operation of sensors such as the heating element, pricking assembly, and water-level controller.

- **Storing data locally** if there is connectivity loss with the central server, and data synchronization with it once connectivity is reestablished.

3. **Local Rule Engine**: The LRE will host ML models that enable ideal boiling conditions and perform analytics on the video feed, received from connected video cameras to determine the egg characteristics as well as the current boiling state. This also helps enforce defined rules (for example, allowing/disallowing the entry of water into the compartment depending on the level indicated by the level sensor, stopping the heating elements if the temperature goes above the allowable threshold, and determining the egg characteristics by combining data from the 2D/3D video cameras and the weight sensor). Almost all the video analytics are done locally to minimize latency and optimize the available bandwidth. The LRE is also responsible for indicating the current boiling status using the attached LEDs.

4. **Central server**: The central server aggregates data from multiple egg boilers and host components such as DT, GRE, AI/ML model creation, and deployment logic. The central server also hosts functionality to deploy firmware upgrades/patches onto the DGs and manage security requirements – for example, certificate rotation. The feedback from users regarding the quality of boiled eggs is analyzed by the central server, and this information is used for model refinement.

5. **Digital twin**: The DT acts as a virtual representation of every egg boiler. The user is able to view the current state of the egg boiler and initiate required actions if they are outside the premises. If there is an egg boiler malfunction, the DT is used to send troubleshooting instructions to the egg boiler.

6. **File upload**: In this example, the file upload pattern is used to enable firmware upgrades to the DG and (in some cases) end devices (sensors and actuators) and to enable certificate rotation.

7. **Device management**: This pattern is used to onboard and deboard DGs and associated end devices, monitor the connectivity status of individual DGs, fetch the system state, and issue troubleshooting commands.

8. **AI/ML integration**: AI/ML models created on the central server and deployed on the DG will help to analyze the boiling conditions that result in the best boiling output. In other words, this component is instrumental in mapping an egg's unique characteristics to its ideal boiling conditions. This is also used to predict any malfunction of the egg boiler by analyzing the operating conditions.

9. **External system integration(s)**: The central server can be integrated with external systems such as payment, **Customer Relationship Management (CRM)**, and **Enterprise Resource Planning (ERP)** systems.

   Integration with a payment system will allow a device manufacturer to enforce the payment conditions/plans with the end user. For example, a *basic plan* will only allow a fixed boiling pattern, whereas a *premium plan* will enable model refinement. This integration will also ensure that the service is provided to only those users who don't have outstanding dues.

   As user feedback is a critical component for the success of the current use case, integration with CRM systems will ensure that feedback is provided by authentic users only; also, it will help to correlate the feedback with historical data to eliminate extraneous feedback. Additional analytics can be done to map feedback to a person's demographics – for example, age, gender, and region.

   An ERP system will provide additional device (boiler) metadata that can be used to determine information such as device location and warranty information.

   At the DG level, there can be local integrations with other smart home devices (an announcement on a smart speaker once the boiling operation is over). Similarly, operational and diagnostics APIs exposed by the smart egg boiler can be consumed by a smart home mobile app (to provide local monitoring/control) that interfaces with other similar smart devices in the home/kitchen, avoiding the need to have a separate app for egg boiler.

10. **Actuator**: This use case relies on multiple actuators to complete its operation:

    - A **water-level controller** is connected over the I2C channel with the DG and controls the amount of water that is available for the boiling operation.

    - A **heating element controller** controls the temperature inside the boiler and is connected over the I2C protocol with the DG.

    - A **pricking assembly**, as shown in *Figure 4.5*, is another form of actuator that is connected with the DG over I2C.

    - LEDs that show the status of the boiler operation are connected over the I2C channel.

The section introduced an innovative idea of the smart egg boiler and how it can be realized using IoT patterns.

Summary

This chapter demonstrated how the IoT patterns introduced in previous chapters can be used to realize interesting use cases in the consumer domain, thereby demonstrating the efficacy of the patterns. This chapter offered a glimpse of the type of sensors and actuators that we need for our use case implementation. Additionally, this chapter looked at the type of architectural decisions we need to make (for example, the type of logic that must be executed at the edge or DG versus the logic that needs to be implemented or hosted in the central server).

The next chapter will continue our journey, where we will see the type of use cases that are relevant in the retail domain and how they can be implemented using the architectural patterns introduced in previous chapters.

5
Pattern Implementation in the Smart City Domain

There is a huge focus on enhancing the infrastructure of existing cities to make it *smarter* and more efficient. Adding sensing and actuation capabilities to parts of a city's infrastructure, such as utilities (lower wastage and minimal disruptions), power plants (safer operations and predictive behavior), traffic systems (for controlling congestion), law enforcement (crime prevention), educational institutions (enhanced engagement for students and automation of regular tasks for teachers), and healthcare services (timely emergency response), and integrating the required communication technologies will make cities safer, more intelligent, environmentally sustainable, and energy-efficient, resulting in improvement in living conditions of city dwellers.

The chapter introduces a few use cases that are relevant to a *smart city* and how these can be implemented by leveraging the IoT patterns described in earlier chapters. Specifically, this chapter provides details about four use cases:

- A smart speaker for modernizing education
- Monitoring the condition of perishable goods
- Driver behavior monitoring
- Automatic replenishment of consumables and raw materials
- Additional use cases

The first use case is innovative and the remaining three are relatively well-known use cases in a smart city context. After going through the chapter, the reader will be able to conceptualize, architect, and implement additional use cases in this domain.

A smart speaker for modernizing education

The capability of smart speakers to recognize voices and their ability to process information at the edge and integrate with central server services can be a potent combination for revolutionizing education and providing a level playing field for all students.

In addition to smart speakers and central server services, a mini solar plant is required to power smart speakers and the related hardware (such as routers) on the school premises, as the electricity supply is generally choppy in developing countries.

This use case can be further subdivided into the following two sub-use cases:

- **Transforming education by providing benefits to students and teachers**: Students gain by improving their verbal communication skills, resulting in enhanced confidence, and teachers stand to gain by automating mundane educational tasks:

  - **Improvement in pronunciation for non-vernacular languages**: An example of this is English in the Indian context. Input to the smart speaker is subjected to a central server-hosted AI/ML engine to compare the received input with reference audio files (converting audio into text and regenerating audio using a text-to-speech engine). The smart speaker then provides a list of wrongly pronounced words.

  - **Filler word reduction**: Text is obtained from the received audio input and filler words (such as *um*, *aah* etc.) in the output text are counted. The smart speaker then lists the filler words, along with their occurrence. This would be a three-step process as given here:

    i.   Listening to students' articulation

    ii. Comparing the expected pronunciation with the recorded speech (central server services for converting audio into text and text into audio would be leveraged)

    iii. Providing feedback regarding mispronounced words and a list or count of filler words at the end of the recording session

  - **Automating attendance administration/viva voce**: The smart speaker would be used to ask questions from a predetermined set and record the responses. The responses can then be analyzed by generating text from the received responses and comparing it with correct answers. Similarly, the attendance process can be automated by recording response from each student.

- **Scene articulation for visually challenged students**: This involves students who are visually impaired requesting specific video content; smart speakers and central server services then work in tandem to generate scene information as metadata. Scene metadata sequenced with audio already present provides vivid scene imagery. The required metadata can be recorded manually using a smart speaker by volunteers by pausing the video, articulating the scene information, and then resuming the video to record the original audio, or in an automated fashion by using central server services specifically designed for voice recognition. In a way, the smart speaker can act as *eyes* for visually impaired students by opening access to educational/edutainment videos that would otherwise have limited utility.

A **Local Rule Engine (LRE)** deployed on a smart speaker and a **Global Rule Engine (GRE)** deployed on a central server would contribute significantly to the education sector in developing countries by doing the following:

- Democratizing the availability of video content for visually impaired students

- Acting as a language learning aid for students

- Putting effective tools in the hands of teachers – enabling them to focus on learning innovative teaching skills, as the amount of time spent on mundane activities is considerably reduced

The features related to the current use case can be implemented by following a platform-centric approach whereby the platform provides the basic capabilities of **voice recognition** (audio-to-text) and **voice synthesis** (text-to-audio).

 Another point worth noting is that the platform should be able to process information at the edge as well as in the central server (the LRE and GRE as stated in the **Rule Engine** pattern (refer to *Chapter 3* for more details). The LRE would be used to process wake-up phrases, as well as to handle simple queries. Complex queries would be sent to the GRE for processing and the results would be sent back to the LRE and would be ultimately communicated to the end user.

An AI/ML pattern would also play a critical role here – primarily for processing audio commands and determining the appropriate response. The system must be self-learning as well as self-correcting since the level of queries is expected to increase both in complexity as well as diversity with each passing day. The GRE would create ML models at regular intervals and push them to the LRE (hosted in the smart speaker) so that the smart speaker could answer the maximum number of queries even without network connectivity.

An implementation of the smart speaker use case and the segregation of responsibilities between the GRE and the LRE are shown in the following figure:

GRE functionalities

1. ML model training/creation for responding to complex queries
2. Providing textual responses to queries sent by the LRE/smart speakers
3. Generation of push notifications
4. Metrics and data collection related with usage scenarios.
5. Reporting/processing etc.

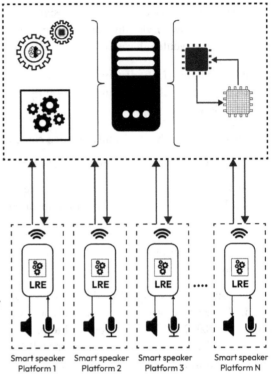

LRE functionalities

1. Processing of "wake up"phrases and responses for simple queries
2. Execution of ML models pushed by the GRE for handling intermittent connectivity issues.
3. Voice analysis (voice-to-text)/ Voice synthesis (text-to-voice)

Figure 5.1 – Realization of the smart speaker use case by leveraging patterns at the edge/local and central server

> **Important note**
> The figure merely illustrates how the LRE and GRE can be used in complementary roles. It does not intend to list all the possible architectural patterns that can be used.

This section described the smart speaker use case and its implementation. In the next section, we will look at another use case involving real-time condition monitoring for perishable goods.

Monitoring the condition of perishable goods

In this use case, a fleet of trucks carries perishable goods from one location to another and the state of the goods being ferried is continuously monitored.

For tracking the state of goods being transported in near real time, a set of sensors (e.g., for temperature, moisture, gas, etc.) will be installed/placed near the goods and will send the current state to the central server along with the current location (the truck's GPS location). The data received is analyzed at the central server and suitable action(s) are relayed back to the specific truck (or to the complete fleet in special scenarios) so that appropriate action can be taken.

Some amount of analysis (and action) needs to be performed locally, especially in scenarios where connectivity with the backend is erratic. For example, if the condition of goods has deteriorated beyond an acceptable limit, it would be prudent to discard the shipment rather than complete the journey to the end destination. Typically, this type of analysis would be performed at the central server; however, as connectivity might not be available throughout the shipment route, these analytics need to be performed at a local/edge server (i.e., at the individual truck level).

There would also be a need by the IoT subsystem to update dependent external systems (part of the supply chain data pipeline) about the current shipment details (delayed/timely delivery, the current condition of goods, etc.) so that downstream workflows can be triggered (for example, ordering a shipment from an alternate vendor if the shipment condition is not good).

Data from the fleet needs to be aggregated at the central server for more extensive analysis and to trigger actions from the insights generated, which necessitates provisioning a rule engine at the central server.

As the connectivity of the fleet to the central server can't be always guaranteed, the state of each truck (including the shipment state) needs to be saved onto the **digital twin** (**DT**) hosted on a central server. The DT would also be used for storing instructions to be passed on to the individual trucks; for example, "*Due to an order change, shipment X needs to be delivered to location Y instead of location Z.*"

The key components within this use case are illustrated in the following figure:

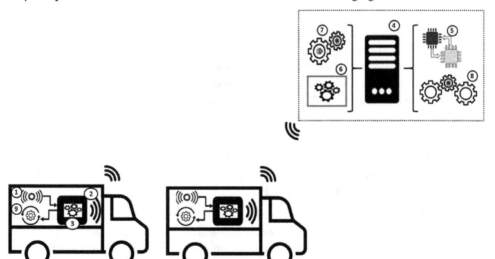

Figure 5.2 – Realization of the condition monitoring use case by leveraging IoT patterns

This use case can be realized by leveraging the patterns mentioned in previous chapters – some of the patterns need to be implemented at the edge (at the individual truck level) and some will be implemented at the central server, as shown in the preceding figure. Details about the patterns leveraged in implementing the use case are elaborated here:

1. **Sensor**: Sensors will monitor physical conditions such as temperature, humidity, pressure, vibration, CO^2 levels, ammonia levels, and so on. The choice of sensors will depend on the material being shipped. Similarly, a GPS sensor will be present to send locational information. These sensors will send the current parameter values to the device gateway via supported interfaces.

2. **Device Gateway** (**DG**): The DG receives data from sensors and sends it to the central server. It is also responsible for buffering data in case of loss of connectivity with the central server. Similarly, it will receive commands/notifications from the central server (e.g., a changing thermostat value, a changed shipment route, etc.). Also, in some cases, the DG will act as a protocol (or data format) translator if sensor/actuator data needs to be converted into a standard format before being sent to the central server.

3. **Local Rule Engine** (**LRE**): The LRE hosted on a DG will evaluate basic rules and analytics, especially in scenarios when the connectivity with the backend is not available (e.g., setting the thermostat to a given value based on the observed ambient temperature). Similarly, some of the rules that need to be evaluated on an almost real-time basis (and hence can't afford the latency involved in a round trip to the GRE/central server) will also be evaluated on the LRE.

4. **Central server**: All the trucks in the fleet will connect to the central server to push data, as well as receive commands. Central server hosting will provide adequate scalability/elasticity in case the number of trucks increases or decreases. As complex analytics are expected to run on this server, generally, compute and storage requirements are much higher here than for the DG.

5. **Digital twin** (**DT**): The DT will act as a virtual representation of each truck in the fleet. Queries raised by the end user to understand the state or condition of an individual truck will be resolved by the DT. Also, any command sent to an individual truck (e.g., to change the route) will again be sent via the DT only. In this manner, the DT acts as an information buffer between trucks and higher-level applications that are hosted within the central server. This will also remove tight coupling between higher-level applications with lower-level concerns (e.g., whether a particular truck is within the connectivity range or not).

6. **Global Rule Engine** (**GRE**): The GRE will execute rules based on events received from the complete fleet. Rule execution will be similar to that done at the LRE but differ in terms of the data volume and algorithmic complexity. In fact, some of the event processing will fall in the domain of **complex event processing** (**CPE**).

7. **AI/ML integration**: AI/ML will help automate some of the decisions that need to be made, such as monitoring calibration drift for sensors/actuators and automating the associated remediation steps, predicting performance bottlenecks and operational failures while eliminating/minimizing false alarms, and so on. The level to which AI/ML is leveraged will primarily depend on the specific implementation and the level of automation that is required.

8. **Enterprise/legacy system integration(s)**: An IoT system will act as a data conduit with existing external systems for sharing fleet data and triggering the required enterprise workflows. In this case, the data that is shared with external systems and invoked workflows will largely depend on the enterprise context. In some scenarios, the IoT system will communicate with external systems to order fresh supplies in case the current shipment is no longer usable. Another possible scenario is in case some organizations monitor the condition of trucks in addition to monitoring the condition of shipments. Additional data (e.g., CAN data) will be sent along with data related to shipment conditions. Events from the IoT system indicating that some of the truck parts need replacement/repair will trigger downstream workflows such as notifying vendors to arrange for defective parts, scheduling an appointment with the nearest service station for truck repairs, and so on.

9. **Actuator**: The actuator will carry out the commands sent by the LRE or GRE. In this case, the actuator will be in the form of a thermostat and the commands will be to set the temperature to a particular value. Again, as can be seen, the actuator can take different forms (depending on the application scenario) but its main function is to convert the virtual decision (taken by the LRE or GRE) into a real-world manifestation.

After understanding how monitoring the condition of perishable goods can be implemented, we will explore another interesting use case about monitoring driver behavior in an automated fashion.

Driver behavior monitoring

A significant number of road fatalities could be averted if a driver's overt/covert behavior is monitored in real time. Overt abnormal behavior such as that related to drowsiness (or driving in an intoxicated state) can be detected by a camera with the ability to analyze video feeds locally at a DG installed in the vehicle. Trained ML models for analyzing video feeds are pushed periodically from a central server. These models gauge and report driver behavior, along with a confidence level. Analysis of behavioral patterns can be further augmented by obtaining additional data from the driver's wearable device (such as pulse rate and last night's sleep quality).

Processing needs to be done locally (at the edge) as video feeds can't be sent to the central server as it would hog the communication channel's bandwidth, as well as the response not being within expected time limits.

Along with the video feed, other crucial parameters (sharp turns, sudden brakes, acceleration patterns, and so on) need to be analyzed and correlated with video analytics to arrive at a more holistic and accurate understanding of a driver's behavior. This additional data would be obtained from an **Engine Control Unit** (**ECU**) and **On-Board Diagnostics** (**OBD**) port, which come pre-installed in almost all modern vehicles.

Once the driver's behavior is accurately determined and is co-related with vehicle key operational parameters (e.g., speed, braking/turning patterns, etc.), interesting use cases can be realized by an

LRE, where events will be analyzed to determine whether a preconfigured rule is violated, and if they are, an associated action can be triggered. Some of the possible actions are listed here:

- Dynamic monitoring and control over the allowed speed limit – as in, the speed of the car can be restricted if divergent driver behavior is observed

- Sending alerts to law enforcement agencies

- Curbing vehicle speed beyond an upper limit

- Blocking entertainment systems to reduce distractions

- Notification to insurance companies for adjusting an insurance premium based on driving history

The key components of this use case are illustrated in the following figure:

GRE functionalities

1. ML model training/creation for detecting driver's divergent behavior (drowsiness and intoxication) using video analytics
2. ML model for determining rash driving conditions (sharp turns, sudden brakes, speed limit violations)
3. Model updates pushed to vehicles on event basis or at a predefined frequency
4. Metrics and data collection related with driving patterns and rule violations
5. Reporting and sending of alerts to law enforcement agencies

LRE functionalities

1. ML model execution for detecting and reporting divergent behavior and rash driving events – input received from installed video camera feeds and the vehicle CAN bus interface
2. Invocking actions based on identified events (send notifications or moderate the speed in case of divergent behavior).
3. Usage of sensing mechanisms, such as wearables on driver, CAN interface, video camera feeds

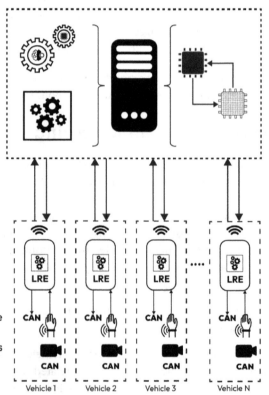

Figure 5.3 – Realization of a driver monitoring use case by leveraging a DT, LRE/GRE, DG, and AI/ML patterns

This section went through the implementation of a driver behavior monitoring use case and its implementation. The next section will detail a scenario in which consumables can be ordered automatically once the available quantity reaches a defined threshold.

Automatic replenishment of consumables and raw materials

Consumer devices and appliances are getting smarter and more connected and we are not far off from the day when they will be able to detect and order the required consumables or raw materials by themselves. These appliances will detect whether they are running low on raw materials/consumables and then notify the user about the consumables that need to be replenished and leave the final purchasing decision to the owner. Over time, the system can learn someone's behavior/preferences and even make the purchasing decision on their own. Appliances will have their own processing logic and will connect to a mobile device (using **Bluetooth Low Energy** (**BLE**) or a similar protocol) to send their statuses to a central server.

The key components of this use case are illustrated in the following figure:

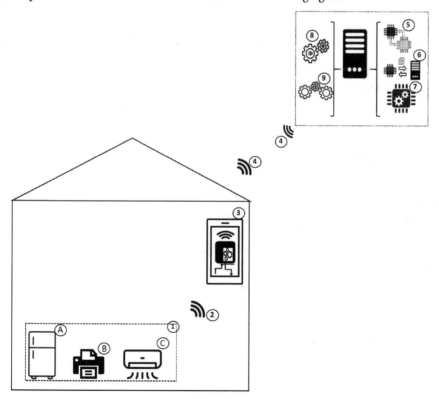

Figure 5.4 – Realization of an automatic consumable ordering use case by leveraging IoT patterns

The main components of an automatic ordering use case are elaborated here:

1. **Consumer appliances**: Various consumer appliances (a coffee maker, blender, toaster, microwave, pressure cooker, kettle, water purifier, vacuum cleaner, air conditioner, oven, dishwasher, printer, and refrigerator) can publish their current consumables/raw material statuses to the user as notifications on their mobile device. The three representative appliances are detailed here:

    A. **Refrigerator**: The refrigerator is equipped with sensors such as liquid level detectors (which detect the amount of milk/water/juice left), and can also use video cameras to determine the amount of remaining fruits and vegetables.

    B. **Printer**: The printer's ink cartridge level can be communicated to a central server via a mobile device so that it can be replaced on time. The printer can also communicate its operational status to a central server, which, in turn, can send repair recommendations to the user.

    C. **Air conditioner**: The air conditioner can either send operational (the filters requiring cleaning) or maintenance (the probability of the compressor malfunctioning in forthcoming weeks) information to the user.

2. **Local connectivity**: Using Wi-Fi or BLE pairing, the smart appliances will connect to the DG and this will enable bi-directional communication with the DG.

3. **Mobile device**: The mobile device here will perform the role of DG and serves two main purposes:

    -   Display the notifications generated by appliances and allow the user to accept the recommendations regarding ordering supplies and authorizing payments

    -   Enable connectivity with the backend server and perform local analytics/process local rule actions

4. **Long-range connectivity**: Long-range connectivity such as Wi-Fi/cellular connectivity is required for a mobile device/DG to connect to the central server.

5. **Digital twin**: The DT will act as a virtual representation of every home. Users will be able to view the current state of their home appliances. A DT for the home can be designed in a hierarchical fashion, where DTs of individual rooms, garages, washrooms, and so on are part of the overall house's DT. The need to have this hierarchical DT, as well as the number of hierarchy levels, will depend on the complexity and scale of the individual house. As mentioned earlier, both the operational as well as maintenance-related information will be available for all the connected devices. Even the connectivity status (connected, not connected, last connectivity status, etc.) will be available.

6. **File upload**: This pattern will be used to enable firmware upgrades to the DG (updating the version of a mobile app, for instance) and (in some cases) to end devices (appliances) and for enabling certificate rotation.

7. **Device management**: This pattern is used to onboard and deboard DGs and associated end devices and monitor the connectivity status of individual DGs. This would also be used to fetch the system state and issue troubleshooting commands.

8. **AI/ML integration**: This pattern can help in predicting the need for maintenance of appliances by analyzing the operational data and comparing it with known failure models. It can also determine the right vendor to purchase supplies given the different combinations of price and delivery timelines offered by vendors. Similarly, it can analyze the customer feedback available on social media channels to avoid purchasing from non-reputable vendors.

9. **External system integration(s)**: This integration will enable the system to determine vendors who can provide supplies within the expected budget and delivery timelines. Similarly, integration will also help to find the right skilled person who can perform preventive maintenance. The system can also determine whether the appliance is within warranty and initiate a service request to prevent appliance malfunction. Integration can leverage users' historical health data and suggest supplies that are conducive to somebody's overall health and vitality.

These sections covered various use cases that can transform a city into a smart city. Let's look at some additional use cases in the next section.

Additional use cases

Some of the additional use cases that can be implemented in a smart city are shown in the following diagram:

Environmental sustainability	E-Governance	Public Safety	Smart Buildings
			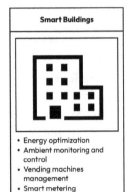
• Leakage prevention • Waste management, waste recycling • Air quality monitoring, weather monitoring • Grid protection, demand forecasting	• Digitization of ownership records • Grievance management • Online payments and reimbursements • Citizen engagement	• Law enforcement • Emergency services • Security and surveillance, crime prevention, geo-fencing intrusion detection systems • Fire safety	• Energy optimization • Ambient monitoring and control • Vending machines management • Smart metering

Transport	Transport	Transport
	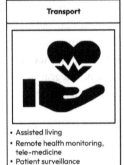	
• Fleet optimization, route optimization • Congestion control • Smart parking • Asset tracking	• Assisted living • Remote health monitoring, tele-medicine • Patient surveillance	• Virtual learning • Interactive content • Automated attendance management

Figure 5.5 – Possible smart city use cases

These applications can be developed to make life of the city dweller more meaningful and interesting.

Summary

The scope of the smart city is much wider than what is covered in this chapter and will touch the lives of city residents in multiple ways.

This chapter demonstrated how IoT patterns can be used to realize some of the use cases. Additionally, the chapter provided details related to implementation nuances (sensors, actuators, connectivity, etc.) and how these change from one use case to another. The list of use cases provided is by no means exhaustive and the reader is encouraged to use the knowledge gained in this chapter to devise solutions that would solve particular problems/issues that are prevalent in their city.

In the next chapter, we will continue this journey and examine some use cases that are relevant in the retail domain and how they can be implemented using architectural patterns.

6

Pattern Implementation in the Retail Domain

The chapter provides an overview of the **retail domain** and how IoT is set to transform this domain by enabling innovative use cases and applications. This chapter will help the reader in understanding the retail domain from a historical perspective and how IoT, along with other technologies, is expected to catapult it to another level.

The chapter also lists the challenges faced by retailers and how these challenges can be effectively mitigated by IoT. The information gained in this chapter is then used to show how a next-generation **retail store** can be realized, where the **store**, the **shopper**, and the **merchandise** are continuously monitored to gather useful insights and improve store operation.

An overview of the retail domain

The term **retail transactions** refers to the transactions where a product is sold directly to an *end consumer* for their own consumption. The retail industry is one of the fastest-growing industries and is marked by intense competition among existing players. Also, shoppers have high bargaining power due to the non-differentiated nature of the goods and the presence of multiple suppliers. All of this makes it extremely difficult for retailers to attract and retain a loyal customer base.

Another interesting aspect is the huge difference of scale at which retailers operate – ranging from tiny neighborhood mom-and-pop stores to all the way to mega stores. Another key feature of this domain is that shopper behavior doesn't change drastically in short term, and this makes it easier for retailers to run analytics on the purchase history and provide more reasonable recommendations.

The domain has seen several transformations since its inception and these changes can be broadly categorized into four major transitions, as shown in the following diagram:

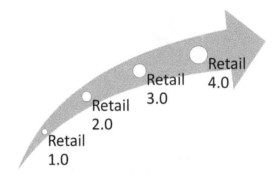

Figure 6.1 – Stages of the retail industry's maturity

The key characteristics of the transitions shown in the previous figure are as follows:

- **Retail 1.0**:
    - Physical (brick-and-mortar) stores
    - Single-channel
    - Self-service
    - Cash transactions

- **Retail 2.0**:
    - Departmental stores and hypermarkets
    - Suburban plazas
    - Credit card transactions
    - Loyalty programs
    - Minimal customer purchase analytics

Retail 3.0:

- E-commerce
- Global supply chains
- Predictive delivery timelines (days)
- Extensive purchase analytics
- Basic recommendations
- Online payments

Retail 4.0:

- Multi- and omni-channel

- IoT, robotics, AI, and **Augmented Reality/Virtual Reality (AR/VR)**

- Social media reviews

- Self-checkout and **buy online, pick up in-store (BOPIS)**

- Showrooming and web-rooming

- Real-time and end-to-end tracking of merchandise

- Recommendations based on customer segmentation and related purchases

- Automated replenishments

- Predictive delivery (hours)

- Hypercustomization of orders

Retailers face a diverse set of challenges, which includes the following:

- Increased complexity in the supply chain, diverse sourcing patterns, and increased variety in sales and distribution channels. This is further exacerbated by intense competition and reduced margins.

- Expectations of omni-channel presence and hyper-personalization on the part of shoppers and the related challenge of keeping inventory levels synchronized between online and physical stores.

- Shoppers expect a seamless transition between channels – for example, online shopping loyalty programs should be automatically carried forward when somebody visits a physical store. Similarly, retailers should be aware of an online purchase that a shopper made few hours earlier and shouldn't recommend the same or similar products but rather complementary or related products.

- A reduced product life cycle and the corresponding pressure of continuous inventory replenishment.

- When cameras based on technologies such as **light detection and ranging (LiDAR)** or **Infrared (IR)** are used to track shopper movement/behavior, the data gathered may be incorrect as these technologies are unable to account for movement of retail staff within the retail store. Although, data can be corrected *out of the band* by excluding staff movement through the provision of unique tags, however, this makes the overall implementation complex. Luxury stores are typically more impacted by this, as the ratio of consumers to staff is quite low.

- With the convenience offered by e-commerce sites, offline/brick-and-mortar stores are facing difficulties in attracting and retaining shopper footfall.

- The inability to gain real-time visibility of key metrics such as inventory levels results in stockouts/overstocking situations.

- There is a need to ensure efficient utilization of energy and packaging materials and adherence to overall sustainability goals. There is influence from governmental agencies here, as well as from youngsters/millennials who want to purchase from environmentally conscious brands.

- Retailers are quite well versed in the nuances of the industry; however, they find it difficult to hire and retain skilled personnel who can integrate diverse technologies such as IoT, Big data, AI, ML, and so on, which are required to implement advanced retail solutions.

This chapter covers IoT applications that are relevant within a retail store. However, it is worth noting that IoT plays a vital role not only within a retail store but also has an impact throughout the supply chain – for example, a retailer can provide a shopper with the accurate availability time of an out-of-stock item by checking the item's inventory levels at the manufacturer site and then factoring in additional variables such as transportation time. Estimation can be further refined by considering external factors such as weather, road conditions, social disturbances, and so on.

Retail goods change multiple hands before they reach a retail store, and the related stages are shown in the following figure:

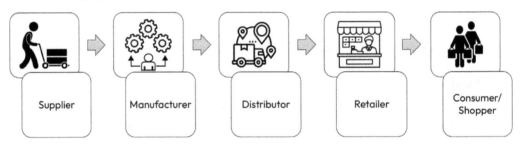

Figure 6.2 – Transfer of retail goods from supplier to shopper

It is beneficial to track the location and condition of an item throughout the supply chain rather than tracking it only when it reaches retail store. This makes delivery timelines more predictable, eliminates waste, and helps in ensuring that the received item is of the expected quality.

Omni-channel presence is the de facto expectation from shoppers. Retailers obtain insights into shopper purchasing behavior by leveraging data from both e-commerce sites and physical stores. Shopper touchpoints can be analyzed for optimizing the shopping experience by tracking metrics and **Key Performance Indicators (KPIs)** such as the following:

- Footfall count

- Dwell time

- Conversion rates

- Inventory turnover

- Physical versus digital traffic

- Customer segmentation

- Product segmentation

- Shopper lifetime sales

- Cross-selling and upselling opportunities

- Demand forecasting

- Shelf and space utilization heatmaps

- Planogram metrics

- Brand/product loyalty

- Customer churn

- Customer decision trees

- Arrival time prediction

Now that we have good background on the retail domain, let us understand how real-time data ingestion can be done and how the data ingested can be leveraged.

Using real-time IoT data

Real time IoT data (clickstream data from websites and footfall data from physical stores) is integrated with other systems, such as ERP, CRM, and **Point-of-Sale (PoS)** systems. This helps generate deeper and broader insights such as the following:

- The number of items on shelves can be determined to ensure automated inventory replenishment, and related systems (financial management, invoicing, taxation, and so on) are kept synchronized for increased efficiency and error reduction.

  In addition to determining the number of products, this monitoring capability can be integrated with a *pricing engine* to recommend discounts on old products.

- Operational (energy consumption) and diagnostics data from retail store equipment (e.g., refrigeration, air conditioning, and lighting equipment) can be used to perform predictive maintenance, ensuring 24/7 operation.

- Sales trends and their correlation with factors such as seasonality help in redefining existing business models and predicting demand patterns and supply fluctuations, and this can be used in determining the effectiveness of past promotions.

- Personalized and targeted promotions, product recommendations, and loyalty programs are achieved by analyzing purchase history and performing detailed customer segmentation. An example of advanced personalized prediction would be a case in which the retailer anticipates the need for a product even before the shopper goes to buy something (by deducing intent from social media comments, for example).

- Product recommendations are arrived at by analyzing both the shopper's own data as well the data of people in similar demographics:

  - **Recommendations based on a shopper's purchasing history**: Here, systems analyze the purchasing history of the shopper to understand product preferences, and products that fit with past purchases are provided as recommendations

  - **Recommendations based on the purchasing history of other shoppers who belong to the same demographic group**: Here, past purchases of other shoppers are analyzed, and products are recommended based on similarity in demographics – for example, age group, gender, ethnic background, economic status, and so on

- By analyzing the time spent by shoppers in various store locations, the optimum physical store layout and staff requirements can be determined, reducing operational costs. This will also help in placing *low-visibility items* in areas with more footfall. Similarly, the ideal product placement can be determined, which can aid in designing planograms, shopper journeys, and so on.

- Wait times at cashier counters can be predicted and managed, including by providing options for self-service, self-checkout, self-search, and self-support, as well as faster *whole basket/whole trolley* checkouts, ultimately moving toward a *cashier-free* vision.

- Optimal pricing strategies are determined by leveraging AI/ML technologies that consider deterministic and non-deterministic factors such as supply chain vulnerabilities, seasonal trends, and competitor pricing.

- A **geo-fence** is created by using technologies such as RFID and Wi-Fi to help prevent retail shrinkage by flagging shoplifting incidents. These technologies can also be used to flag products that are placed in non-designated locations.

- Retail processes (e.g., return processing) are automated by making the process faster and error-free. Unique identifiers (bar codes, QR codes, or RFID tags) attached to inventory items help update the current store's inventory levels in case of product returns and simultaneously update the systems such as ERP systems. This ensures that physical and virtual inventory levels are accurate and are *in sync*.

- These tags will enable the tracking of inventory items during their entire life cycles – from the time item is received at the store gates to when it is finally handed over to the shopper:

- **Item receiving**: As RFID tags don't have *line-of-sight* requirements, individual scanning of items is not needed and they can be read in groups – making the process fast and devoid of any manual/oversight errors.

- **Locating items**: A combination of RFID tags and readers are used to locate items within a store by using the RFID readers to measure the signal strength emitted by tags – the closer the item is from the reader, the higher the detected signal strength is, and vice versa. The same concept can be used to define geo-fencing to ensure that a particular item does not leave the designated area. This use case can be further extended to automatically ordering items if the count is below the threshold.

- **Checkout processing**: As multiple items can be scanned in one go, this makes the checkout process efficient.

- The attendance of store employees can be automated by adding additional biometric systems such as fingerprint sensors and facial recognition cameras. The attendance data can be further integrated with payroll systems for accurate payment to employees.

- Customer support is made more efficient by ensuring quicker responses to shopper queries and an improved shopping experience.

- By attaching unique identification tags to the inventory items and comparing the available count with the data shown in ERP systems, inventory reconciliation process is automated and performed in a real-time fashion.

- Technologies such as AR/VR can provide immersive and enriching shopping experiences. Some of the possibilities are as follows:

  - Helping shoppers to try various options before purchasing, increasing the store's footfall and customer engagement and reducing the return rate

  - Superimposing additional information (price, availability, reviews, color options, etc.) onto the product view

  - Guiding the shopper to the relevant product and its placement within the store

The following figure summarizes the business benefits of implementing IoT in retail stores:

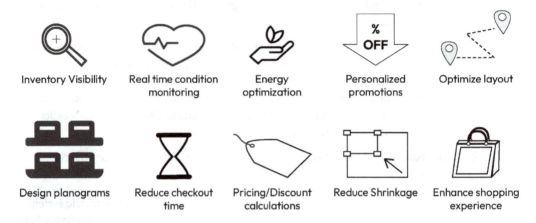

Figure 6.3 – Business benefits of implementing IoT in retail stores

Data from different sources (clickstream data, social media feeds, etc.) can be integrated with other enterprise systems such as ERP and CRM systems to derive useful insights, as shown in the following figure:

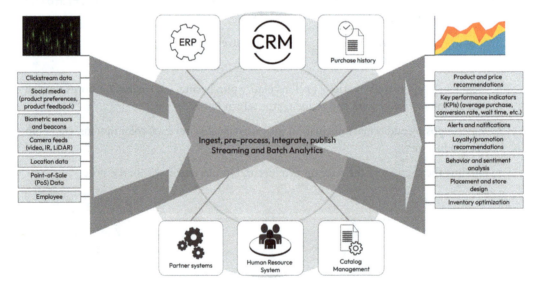

Figure 6.4 – Raw data ingested from multiple sources turning into useful insights

Now that we've gathered a high-level understanding of the retail domain, we can shift our attention to the type of sensors and actuators that are deployed in retail stores to support IoT use cases.

Implementing sensors and actuators in retail stores

The typical kinds of sensors and actuators deployed in retail stores are shown in the following figure:

Figure 6.5 – Typical sensors and actuators used in retail stores

Let's look at each of the elements shown in the preceding figure:

- **Video camera**: This discerns shoppers' attributes (demographics, age group, and gender) and relates them to shopping behavior patterns. This helps generate more targeted promotional content and determine the efficacy of previous promotions. It also eliminates theft prevention. However, it can be used only if shoppers don't have privacy concerns.

- **Bluetooth beacons**: This allows you to send context-sensitive push notifications (such as discounts on groceries) to registered shoppers via a store app.

- **Wi-Fi**: Wi-Fi routers detect a shopper's location (using smartphone Wi-Fi) inside the store and this location information is used to optimize inventory management, calculate dwell times, and display customized or personalized messages, signage, and promotions.

- **RFID**: This is used to uniquely identify products in store and the technology doesn't require line of sight (in contrast with other technologies such as bar codes). Moreover, multiple RFID tags can be read with a single scan. Another advantage is that additional information can be written into RFID tags, such as expiry dates, prices, and so on. These tags are typically used to prevent shoplifting by placing RFID readers near the store exits, which are integrated with an alarm system.

- **IR/LiDAR cameras**: These cameras provide similar functionality to video cameras; however, they help protect the shopper's privacy.

- **Digital signage**: This displays targeted promotions and product information (applicable discounts, alternate designs, available sizes, accessories, and so on)

- **Temperature/humidity sensor and air quality sensor**: These are used to ensure that the right products are stored in optimum ambient conditions.

- **Bar code readers**: Used at PoS terminals and cashier counters for billing.

Now, let's take a look at this use case in greater detail.

Use case – real-time tracking in retail outlets

Tracking shoppers within a retail outlet helps to understand shopper's *objects of interest* and distribute the inventory and staff accordingly. Without real-time tracking of shoppers using cameras, it would be difficult to analyze the store pathways preferred by shoppers. Determining the queue length at cash counters and leveraging past data to predict waiting times helps avoid checkout delays and improves shoppers' experiences.

Privacy concerns

Tracking shoppers can be done by performing video analytics at the edge (using a DG); however, this might result in **privacy concerns**. Retailers need to tread a fine line between gathering data for understanding shopper behavior/preferences and alleviating privacy concerns. Some of the mechanisms that retailers use to ensure this balance are as follows:

- Explicitly informing (and taking consent from) shoppers about the type of data being collated and its intended purpose. Additionally, shoppers should have the option to revoke their consent at any time.

- Consumer tracking can be done by leveraging alternate technologies such as IR, LiDAR, or BLE, which provides (out of the box) shopper de-identification/anonymization. A related technique uses invasive sensor technologies (such as video cameras) but apply de-identification/ anonymization algorithms (e.g., data hashing) before data processing/usage.

- Leveraging IoT security capabilities (as detailed in *Chapter 11*) to ensure that the captured data is not inadvertently or maliciously used.

- Educating/training store employees regarding the importance of protecting shopper privacy and the implications of not adhering to it, as well as specific regulations such as the **General Data Protection Regulation (GDPR)**.

Now, let's take a look at how this use case can be realized.

Leveraging IoT patterns

Like shopper tracking, tracking RFID/BLE-tagged merchandise provides benefits such as real-time inventory updates and pilferage reduction and can help you locate missing/misplaced items. This also enables automation of inbound processes (receiving orders, unloading and storing items, and the general management of incoming supplies).

The key components of this use case are illustrated in the following figure:

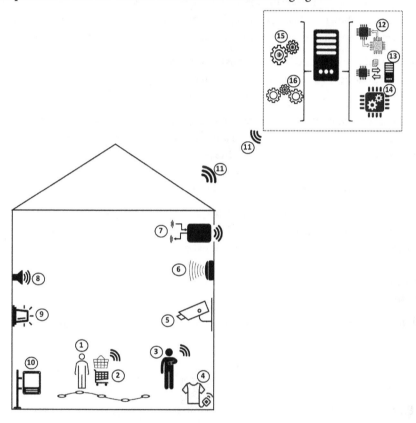

Figure 6.6 – Realization of a shopper/merchandise tracking use case by leveraging IoT patterns

Let's look at these components of the use case in greater detail:

1. **Shopper**: Shoppers' movements can be monitored by analyzing the pathways traversed and the related dwell times (the time spent at certain touch points) using a variety of technology:

   A. Presence in a particular store area can be determined by analyzing a video feed or footage from LiDAR cameras (to mitigate privacy concerns). Video feeds can provide additional demographic data about the gender, age group, and so on of the shopper.

B. The shopper can be tracked using a mobile application downloaded by shopper. The application uses Wi-Fi or BLE technology to determine the location of the shopper's mobile phone and, by extension, the shopper's location. The current location of a mobile phone is calculated by determining the signal strength between the mobile device and a fixed Wi-Fi/BLE source (e.g., a router). For better accuracy, three sources are required to do this, which is known as the triangulation technique.

2. **RFID-tagged shopping trolley or cart**: Trolleys or carts are affixed with RFID tags to determine their current location (for asset tracking purposes) or to determine the path taken by a shopper if a camera is not available, as their movement closely mimics the shopper's movement.

Once special case that needs to be considered is a scenario in which store employees are moving carts/trolleys for cleaning or consolidation purposes. This can be corrected by using the fingerprint/handprint sensor to exclude store employee-initiated movements.

3. **Store employees**: With LiDAR and IR cameras, shopper movement data is skewed by the movement of store employees. As a result, there is a need to find an alternate mechanism to segregate shopper and employee movement data. One possible implementation is a mechanism by which store employees carry RFID tags (attached to their identification cards, for example). Camera data then needs to be time-synchronized with RFID reader data to generate correct analytics.

Time synchronization between multiple devices or sensors is required in many IoT solutions to enable proper ordering of events and accurate coordination of processes and workflows across multiple devices and for non-operational or diagnostics purposes, such as determining the exact flow of events from available logs. Techniques such as periodic synchronization with a **Network Time Protocol (NTP)** server and generating timestamps from a common source are used to maintain time synchronization across multiple devices.

4. **RFID/bar code-tagged merchandise**: Affixing RFID tags/bar codes to merchandise helps in two ways:

A. **Pilferage elimination**: RFID readers placed at the store exit gates generate an alarm if unauthorized merchandise is taken outside of the store premises.

B. **Faster checkout**: Tagged merchandize is invoiced quickly and this is especially relevant for RFID-tagged merchandise as multiple items are read in a single scan.

C. **Self-service/self-support**: The shopper brings the item of interest near to the reader and detailed product information is displayed on digital signage.

5. **Video/LiDAR/IR cameras**: In addition to serving the conventional need for surveillance, these cameras can be used to determine shopper store journeys/paths. By determining the dwell time in different store areas, the layout that ensures higher footfall is determined.

6. **RFID readers**: These are used to detect the presence of RFID tags in the vicinity. The relative signal strength of the detected tag gives an indication of the distance between the reader and the tag, thereby indicating the location of RFID tagged asset.

7. **Device Gateway**: The DG acts as a bridge between sensors and actuators and executes a static set of rules (e.g., if theft is detected, a buzzer is sounded and/or flashing lights are turned on). The feed from cameras will be analyzed at the DG and the analysis results will be sent to the central server for detailed/complex analytics. The sensors and actuators that are attached to the DG in the current use case are shown in the following figure:

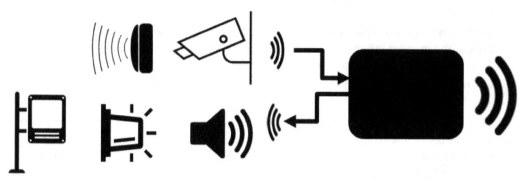

Figure 6.7 – Sensors and actuators attached to DG

8. **Speaker**: This would be used to sound an alarm in case of emergency as well as when theft is detected using RFID readers.

9. **Flashlights**: These would be used in conjunction with a speaker to visually highlight any emergency or theft.

10. **Digital signage**: Digital signage serves multiple functions within a retail store:

   A. Guides shoppers within the store

   B. Increases shoppers' engagement by displaying relevant and context-dependent information, such as product reviews, approximate waiting times, product catalogs, discounts, and so on

11. **Wi-Fi/cellular connectivity**: This is required for a mobile device/DG to connect to the central server.

12. **Digital twin (DT)**: A DT acts as a virtual representation of every store. The retailer will be able to determine the ambient and storage conditions.

   The DT also maintains a reflection of other aspects of the store, such as shopper journeys, employee presence, queue length, and so on. This information is helpful in simulating various *what-if* scenarios – whether the placement of items on a particular shelf increases store footfall and for correlating the impact of discounts on an increase in sales.

   Additionally, the DT holds the current state of all sensors and actuators. Although the content displayed on digital signage is controlled locally, however, in some cases, content can be pushed from a central server by first configuring it in the DT.

13. **File upload**: The file upload pattern is used to enable firmware upgrades to the DG and to end devices (sensors and actuators) and for enabling certificate rotation. Certificates are installed in field devices for authentication purposes and certificate rotation (replacing an old certificate with a new certificate) is required for reasons such as expiry, changes in a device's permissions or authorization, the occurrence of a security breach, and older certificates no longer being trustworthy. This is further discussed in *Chapter 11*.

14. **Device management**: This pattern is used to onboard and deboard DGs and associated end devices and monitor the connectivity status of individual DGs. This is also used to fetch the system state and issue troubleshooting commands. For identification of RFID, QR, or bar codes, these codes are initially mapped to specific items where a unique identification number is associated with an item's metadata.

15. **AI/ML integration**: This pattern helps in predicting the need to maintain various equipment used in store by analyzing the operational data and comparing it with known failure models. It also determines the right vendor to purchase supplies from given the different combinations of price and delivery timelines offered. It also analyzes the shopper feedback available on social media channels to avoid purchasing from non-reputable vendors. The integration also aids in deciding various pricing/discounting strategies.

    This integration helps in analyzing shopper paths/journeys within a store and suggest optimal layouts (or planograms) to maximize a shopper's experience and sales.

    Although the feed from store cameras will be analyzed locally (on a DG), the corresponding model, however, will be created initially by leveraging AI/ML technologies at the central server, and the created model deployed on a DG.

16. **External system integration(s)**: This integration enables systems to determine vendors who can provide supplies within the expected budget and delivery timelines. This also helps to find the person with the right skills who can perform preventive maintenance and determine whether store equipment is within its warranty and initiate a service request to prevent equipment malfunctioning.

    External systems such as HR systems are integrated to automate payroll processing based on the actual time spent by store employees in store. HR system integration helps in keeping RFID tag database in sync and avoids discrepancy due to employees joining/leaving. Integration is also required with third-party advertising companies for displaying promotional content on digital signage.

    Integration with systems such as ERP systems helps to determine whether the inventory level of an item is below threshold to generate a timely purchase order. Integration with an external system is required to gauge the price offered by competitors in real time so that the appropriate price adjustments are made on time.

The section provided details about how some retail domain use cases are implemented using IoT patterns; this brings us to the end of this chapter.

Summary

This chapter provided an overview of retail domain, the challenges faced by retailers, and how IoT can be used to transform this domain. By affixing sensors to humans and equipment, important actionable insights can be generated. Implementing data-driven recommendations enhances the customer experience and helps boost sales.

The chapter also illustrated how IoT technologies help track shopper journeys and understand purchasing behavior that is used by retailers to fine-tune marketing campaigns, resulting in an overall positive experience for the shopper.

The next chapter will start with an overview of the manufacturing domain and then discuss the role of IoT in realizing the vision of smart manufacturing.

Summary

7

Pattern Implementation in the Manufacturing Domain

One of the domains where IoT has contributed (and is expected to contribute further) significantly to digitalization efforts is the **manufacturing domain**. The prime reason is that this domain stands to gain the most from obtaining real-time visibility into manufacturing operations, identifying optimization opportunities, and automating the existing manual processes (e.g., automating the inspection of goods by analyzing the video feed as it moves over the assembly), resulting in increased operational efficiency.

Most manufacturing plants already deploy automation to a certain extent (e.g., **Computer Numerical Control** or **CNC**). Machines are used to perform repetitive tasks such as welding, milling, cutting, and so on by programming a series of predefined instructions. However, there is still huge potential that can be tapped into by deploying IoT technologies end to end (aggregating/analyzing data from all plants and the complete supply chain). Manufacturers view IoT and related technologies as a tool that will enable them to sell not only finished goods but also associated services. Another key reason that the momentum of the adoption of digital transformation initiatives (and IoT technologies) has increased is the risk of complete business disruption caused by low-cost manufacturing hubs.

This chapter aims to provide a historical perspective of the manufacturing domain and explore how IoT is expected to transform it by offering multiple use cases and services. Leveraging IoT in manufacturing is known by different names, whether that be smart manufacturing, **Industrial Internet of Things (IIoT)**, connected factories, smart factories, Industry 4.0, digital manufacturing, and so on. We will be using the term **smart manufacturing**. The chapter covers the following main topics:

- An overview of smart manufacturing (including the key definitions used in this domain)

- Exploring the evolution of the domain

- Realization of a smart manufacturing use case – the automatic inspection of finished goods

An overview of smart manufacturing

IoT plays a foundational role in enabling smart manufacturing; however, there are a few additional or complementary technologies (which we will discuss later in this chapter) that play an equally significant role, as shown in the following figure:

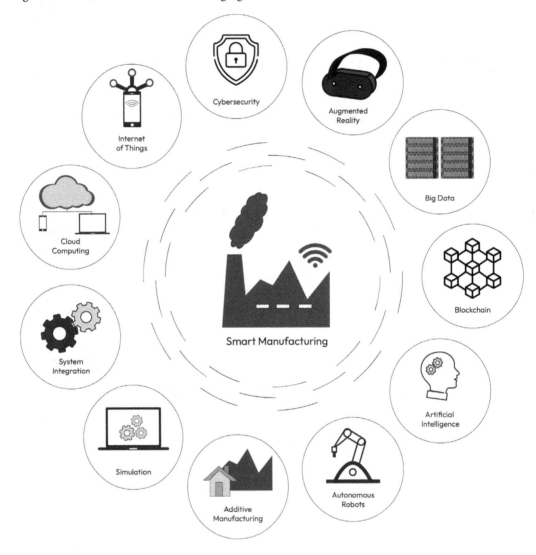

Figure 7.1 – IoT and related technologies that are smart-manufacturing enablers

Readers must be aware of certain terms that are used frequently in traditional and smart manufacturing. Let's discuss these in detail.

Key terms/definitions

In this section, we'll discuss some of the key concepts in the smart manufacturing domain:

- **Digital threads**: Digital threads connect disparate manufacturing processes or systems and provide an integrated view as they traverse through the different stages of their life cycles (conceptualization, production, usage, service, repair, and decommission). A digital thread is enabled by installing sensors across the manufacturing line, as well as within the smart or connected product itself. Digital threads are the foundational components that enable the DT pattern and some of the benefits of realizing digital threads are as follows:

  - Improves quality and yield by comparing the production lots against the specifications

  - Reduces material wastage and optimizes production timelines

  - Increases customer satisfaction through timely resolution of service issues

  - Continuously refines products by analyzing usage patterns

  - Enables collaboration and eliminates information silos

  - Empowers manufacturers to generate additional revenue streams by establishing innovative business models (services along with products)

  Digital threads and their association with the full life cycle of a product can be understood more thoroughly through reference to the following figure:

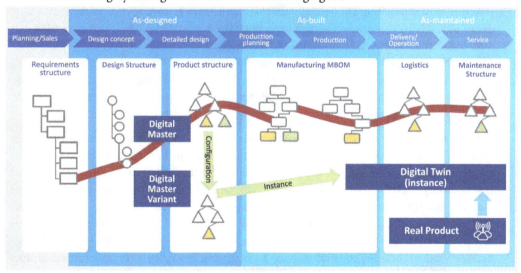

Figure 7.2 – A digital thread encompassing different stages of manufacturing

- **Overall Equipment Effectiveness (OEE)**: OEE helps to measure the current/actual productivity/utilization of the manufacturing plant in relation to its full potential. It is expressed in % and indicates the amount of time for which the manufacturing plant was productive. One of the aspects that makes this metric indispensable in a smart manufacturing context is that it can be measured and reported at any level (at the assembly line, in a department, in a manufacturing plant, and so on) and can be used to objectively compare the performance of, say, one manufacturing plant with another. The formula used to calculate OEE is as follows:

$$OEE = A * P * Q$$

where A is availability (the ratio of the actual production time to the planned production time), P is performance (the ratio of the actual work speed to the planned work speed) and Q is quality (the ratio of the number of actual units produced to the total number planned).

- **Operational Technology (OT)**: OT refers to the automation technology used in manufacturing facilities for monitoring and controlling manufacturing processes. OT refers to a bouquet of technology such as **Programmable Logic Controllers (PLC)**, **Supervisory Control And Data Acquisition Systems (SCADA)**, **CNC** systems, **Distributed Control Systems (DCS)**, and so on. The term is used to differentiate OT from the **Information Technology (IT)** that is typically used to process data generated by OT systems. OT systems are generally deployed on a distinct network and rely on different technologies than those used for IT systems. As a result, the mechanisms of protecting assets from security attacks are significantly different for OT than for IT systems.

- **Discrete manufacturing**: Discrete manufacturing involves the production of distinct items or individual parts of a product (examples include the manufacturing of toys, machine parts, nuts, bolts, smartphones, and furniture) and its key characteristic is that both the raw material and the finished product are countable. Due to the distinctive nature of the output units, these can be tracked during their entire life cycle (provided they are equipped with the required sensors enabled for connectivity) – from the moment they leave a manufacturing facility to final disposal/decommissioning.

- **Process manufacturing**: Process manufacturing, in contrast to discrete manufacturing, involves processing raw materials that can't be uniquely identified/counted and is characterized by the fact that the finished product is manufactured by combining raw materials (or, more appropriately, ingredients) by using specific formulae or recipes. The process may include material transformations induced by a combination of thermal, chemical, and electrical processes. Another key characteristic that differentiates process manufacturing from discrete manufacturing is that in the case of process manufacturing, it is not possible to recycle the finished products or break them down into raw materials. Examples of process manufacturing include the production of petrochemicals and paints. The difference between discrete and process manufacturing is illustrated in the following figure:

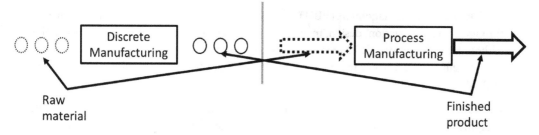

Figure 7.3 – Difference between discrete manufacturing and process manufacturing

- **Manufacturing Execution System (MES)**: An MES is a software system that monitors, tracks, controls, and synchronizes the execution of the manufacturing process, helps to schedule the plant personnel tasks with the objective of providing end-to-end visibility, and provides benefits such as production optimization and supplier/inventory management. Although, there is some functional overlap between an MES and IoT-enabled smart manufacturing (which is the major focus of this chapter), they have significant differences. An MES typically represents a monolithic, rigid, legacy system that contrasts with modular, scalable, and functionally superior smart manufacturing systems. Most manufacturing industries have either transitioned their old MES systems into IoT-enabled smart manufacturing processes and solutions or are in the process of doing so.

- **SCADA**: SCADA is the system used to monitor and control industrial processes primarily from local and sometimes remote locations as well. A user interface, called an HMI in industry parlance, is generally provided for visualizing the operations in real time or near real time and initiating any corrective/preventive measures.

  There is a functional overlap between SCADA and smart manufacturing; however, the latter is more evolved in terms of the technologies used (e.g., **Augmented Reality/Virtual Reality (AR/VR)** and cloud computing), its high reliance on data analytics for generating data-driven insights, and its ability to enrich data by integrating with other enterprise systems (such as **Enterprise Resource Planning** or **ERP** systems, **Human Resource Management Systems** or **HRMSes**, and **Supply Chain Management** or **SCM** systems). In general, SCADA can be considered as a subset of smart manufacturing from a capability standpoint as well as in terms of the value that can be accrued from it.

- **3D printing/additive manufacturing**: 3D printing/additive manufacturing involves the creation of physical or 3D products by progressively adding layers of raw material. It is in direct contrast to traditional subtractive manufacturing, where parts are produced by removing unwanted material from metal blocks through operations such as cutting, boring, and grinding. 3D printing, along with other technologies such as AR/VR, complements IoT in realizing the smart manufacturing vision. It also helps to create quick and cheaper prototypes and personalized products and reduce overall wastage.

- **Product life cycle management (PLM):** This is the process of managing a product's life cycle through conceptualization, design, manufacturing, sales, service, and final decommissioning. Although the definition of PLM may sound similar to that of a digital thread, a digital thread is more refined and broader in scope. In fact, digital threads can be considered the next stage of PLM's evolution.

- **AR/VR:** AR is normally enabled by smart glasses and helps to augment the live view of the equipment or assembly line with additional information – for example, a worker can see the operations of and diagnostics information for a particular machine superimposed onto the actual view of the machine. AR also speeds up maintenance work, as service personnel can view the repair instructions along with an actual view of a machine.

VR differs from AR in the sense that it completely immerses the viewer into an alternate view and doesn't include the live feed component. It is normally used to train service personnel by simulating the actual operations or to generate a simulated view of the product being designed. Both AR and VR can be considered immersive visualization data on top of the data accumulated by IoT sensors and are an interesting way of presenting data compared to traditional paper-based reports. These technologies also enable industrial workers to collaborate remotely while designing new products or iterating over existing designs.

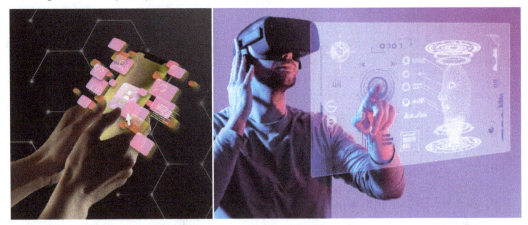

Figure 7.4 – AR and VR in smart manufacturing

Having understood the key terms used in the manufacturing domain, let us understand how the domain has evolved over the years and what advancements can be expected in the future.

Exploring the evolution of the manufacturing domain

The evolution of manufacturing to smart manufacturing and beyond is categorized into five industrial revolutions, as shown in the following figure:

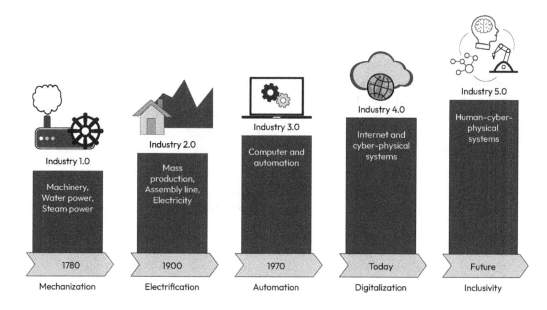

Figure 7.5 – Evolution of smart manufacturing

These revolutions represent significant leaps in terms of productivity and working conditions and are also characterized by the usage of different core technologies. The different stages of evolution are detailed as follows:

- **Industry 1.0**: The first industrial revolution ushered in an era of mechanization and steam engines whereby activities that were earlier performed using human labor started to be performed using machines.

- **Industry 2.0**: The second revolution involved the usage of electricity to power industrial machines.

- **Industry 3.0**: The third industrial revolution relied on using electronics and computers for the automation of production tasks.

- **Industry 4.0**: Currently, we are in the *Industry 4.0* era, where technologies such as sensors, robotics, AI, and the cloud provide additional benefits over Industry 3.0. This stage is detailed later in this chapter.

Industry 5.0: After Industry 4.0 was conceptualized, it was felt that it focused primarily on optimization and efficiency improvements; however, larger issues related to personal, societal, and environmental needs were not given adequate importance. As a result, Industry 5.0 proposes a more human-centric approach where the focus shifts from serving customers to serving employees and includes concepts such as corporate responsibility, sustainability, leveraging human creativity and potential, human-machine collaboration, and hyper-personalized products. Hyper-personalized products not only include products such as automobiles that are specifically customized to meet the end user's preferences but include products from industries such as pharmaceuticals where the products (e.g., pills, vials, etc.) need to be tailored to the specific biological, physical, and psychological needs of the patient.

Another societal phenomenon that makes Industry 5.0 important is the fact that excessive automation is expected to result in unemployment, along with other undesirable socioeconomic effects. The core idea of Industry 5.0 is that both machines and humans possess complementary skills (e.g., machines can perform repetitive or routine tasks efficiently while humans excel in skills such as emotional intelligence, creativity, and empathy) and both can work collaboratively to solve complex problems.

It is also felt that focusing solely on optimization and improving the efficiency of manufacturing plants without giving due consideration to the environmental effects of this is fraught with danger. This, coupled with the fact that, in the future, scrutiny from regulatory bodies and environmental groups is going to increase, was another reason that **sustainability** was added as one of the essential pillars of Industry 5.0.

Now that we've learned about the evolution of this domain, let's look at some of the benefits of this digital transformation.

The benefits of smart manufacturing

The manufacturing industry stands to gain a lot by embarking on a digital transformation journey supported by IoT and related technologies. Some of the key benefits include the following:

- **Ability to sell services along with the products**: Downward pressure on margins, along with the emergence of low-cost manufacturing hubs, is a major driver for manufacturers to find ways of generating additional revenue streams. This will benefit not only manufacturers but also consumers.

  Manufacturers stand to gain an additional customer base, as well as retain the loyalty of existing customers and charge a premium on differentiated services. Consumers stand to gain as they can opt for a utility-based billing model as they pay for the availed service rather than purchasing discrete units (for example, "a temperature maintained at 25° C for the complete month in all the inspection areas" rather than "15 air conditioning units") with the customer relieved of equipment maintenance and upkeep. The manufacturer supports this model by providing connectivity options within the products, which start emitting status/operational information once it leaves the manufacturing plant.

- **Mitigate impacts of retiring or unskilled workforce and enhanced worker safety**: The manufacturing industry is most impacted by an aging or retiring workforce not being replenished at the same rate. Also, this retiring workforce is taking most of its know-how along with it.

  IoT (which can be considered a combination of IT and OT technologies) augmented by complementary technologies (robotics and AR) can help automate most of the manual processes and help minimize the impact of a retiring workforce and help perform almost real-time monitoring of the conditions of manufacturing assets. For example, attaching a vibrating sensor or a sound sensor to a motor and analyzing the data accumulated in near real time by the appropriate AI/ML algorithms gives the same (or even a better) result as that reported by experienced service personnel sensing equipment's condition by sensing the sound/vibration emanating from the equipment. Enhanced worker safety is enabled by geo-fencing hazardous areas, an accelerated/automated emergency response is provided, and workers' exertion levels are monitored in real time.

- **Circular economy enabler**: Like other industries, manufacturing companies are under tremendous pressure to comply with environmental/sustainability regulations. There is a need to optimize energy consumption, reduce operational waste, and design processes that are required by a circular economy.

- **Efficiently respond to dynamically changing market needs**: Going forward, there will be an increased demand for personalized products. This, along with other factors such as supply chain constraints, could make it difficult to accurately gauge the market demands. This type of scenario can be effectively addressed by making production processes more agile – leveraging smart manufacturing tools and techniques. Technologies such as 3D printing can match products to individual preferences and pivot faster to alternate product lines. Tracking of real-time inventory (both at rest and in transit) is enabled by tagging the material along a digital thread.

- **Near real-time visibility into an end-to-end supply chain**: IoT enables visibility into manufacturing operations by reporting metrics such as OEE. OEE, defined as the product of availability, performance, and quality, helps to determine whether a manufacturing plant is running with minimal downtime (*availability*), with minimal waste/discarded products (*quality*), and at full capacity (*performance*). IoT helps to calculate the value of OEE in real time and gives an important insight into a plant's operations.

  OEE provides insights into a plant's yield/throughput. However, further benefits can be accrued by monitoring the complete supply chain (the supply of raw material, goods in transit, delivery to customers, etc.). Tracking the complete supply chain helps to determine any upstream supply chain issues and alter the production plans to mitigate adverse impacts on the plant's output. Within the manufacturing plant, machine vision systems can assist in automated quality control. Additionally, shop floor workers can receive notifications on their mobile devices for possible operational bottlenecks.

- **DTs enhancing product life cycle and production processes**: DTs (refer to the *Digital twins* section in *Chapter 2* for more details) play a vital role in optimizing the product life cycle (from design to decommissioning). DTs help in the design/prototype phase by evaluating different design options, understanding potential constraints/errors, simulating the operation of a product, and gauging the effects of different environmental conditions. All this helps to reduce the cost and timelines of product development, as multiple iterations of the product can be simulated without the need to create a physical product.

The relationship between a physical asset and the corresponding DT is represented in the following figure:

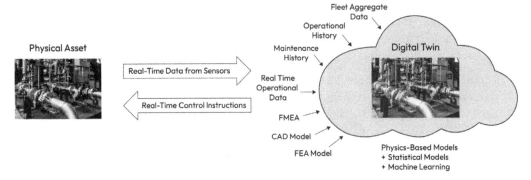

Figure 7.6 – Data communication between a physical asset and DT

During the production stage, DTs help monitor and optimize production processes and determine the optimization opportunities. It monitors the process against the desired behavior, and in case of deviation, raises timely notifications and minimizes variations. This helps to answer questions such as "Is the available quantity of paint sufficient to complete the planned batches?", "Are there any bottlenecks in the upstream processes that will impact the current production activity?", and "Can production processes be altered to minimize the impact of the anticipated bottlenecks?" Essentially, the DT will help to compare the planned process with the actual results and assist plant supervisors to make timely interventions.

To simulate these production processes, data needs to be collated from production machines. This may include the number of units processed, the operational conditions (observed noise levels, vibrational patterns, etc.), as well as auxiliary systems such as **Material Handling Equipment** (**MHE**), including forklifts, conveyor systems, and cranes. Accordingly, the different sensors need to be installed at the desired production stages.

Data needs to be aggregated both locally (or at the plant level), as well as at the central or company level, as this allows for both local decision-making and optimization and global decision-making. Both these are required – local optimization will help optimize the operations of the plant, but central decision-making will ensure that such local optimization is not at the cost of degrading any global operation or sub-optimalization. This will also require employees on the shop floor to have some understanding of IT processes and technologies.

Even the workers working alongside robots can be a source of data that can be fed into the DT to create a more accurate model of industrial operations (for example, the presence of unexpected odors and sounds as detected and reported by industrial workers). Going forward, the DT will segregate responsibilities between humans and robots to take advantage of unique skills and competencies and reduce potential conflicts.

In fact, manufactured goods would continue to report their internal state to the DT even after leaving the manufacturing plant. Smart products leaving the manufacturing plants would be equipped with sensors, which would help to feed the required data (e.g., diagnostics data, operating conditions, etc.) to the DT. This would help to quickly service faulty products, as well as to understand the product usage characteristics that will in turn help refine future product lines.

Hence, DTs can help create a virtuous cycle where feedback from pre-production, production, and postproduction is continuously analyzed and is being used for both process and product refinement, as shown in the following figure:

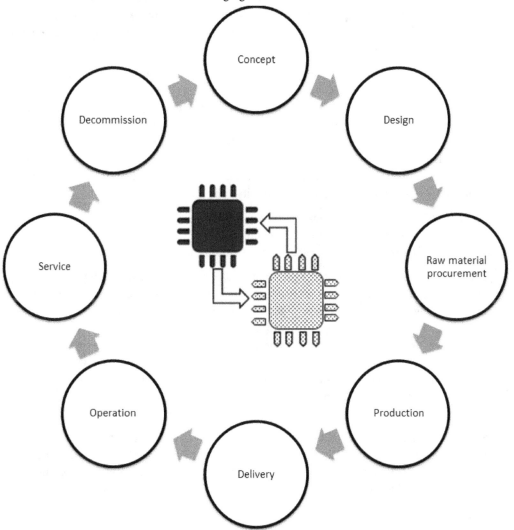

Figure 7.7 – A DT empowering continuous product and process refinements

Despite the obvious benefits of this digital transformation, there have been several challenges in implementing this transition fully. Let's look at these in detail in the next section.

Challenges in transitioning from traditional to smart manufacturing

The transition from traditional to smart manufacturing is not a simple endeavor and involves multiple challenges, some of which are as follows:

- **Inaccessible data**: Legacy machines are not connected and there is no way to get operational data from these machines. Most of the software used in the industrial setup can't send the data outside a plant's parameters. Another prominent issue that hinders the ability to fetch data is the legacy communication protocols that are in use.

- **Data diversity**: Analysis of the unstructured (video or audio data), semi-structured (logs, historian data), and structured (sensor feeds) data generated in industrial plants is difficult. Data corruption, data loss, and data duplication are other factors that lead to complexity. Although these factors are not unique to the manufacturing domain, their effects are much more prominent in the manufacturing, domain as even a minor inaccuracy or anomaly gets amplified and can impact a large number of products before the issue is finally detected and/or rectified. In some cases, this may even require product recall, which is a logistical and operational nightmare for manufacturers and can also impact brand loyalty and reputation.

- **Security**: With data being communicated across a plant's boundaries, the threat of data theft or manipulation increases considerably. The traditional approach of *security by obscurity* is no longer an option, as data needs to be aggregated across different plants for more holistic decision-making.

- **High upfront cost**: The use of sensing technologies, *bolt-on* connectivity for legacy machines, 3D printers, robotics, and so on increases the manufacturing cost and can stall the smart manufacturing journey. Also, it is very difficult to convince the relevant stakeholders to approve the budget as the **Return on Investment** (**RoI**) is difficult to deduce. However, following a piecemeal approach where adoption is done as per a staggered implementation roadmap with well-defined milestones is a preferable approach – for example, retrofitting legacy machines with sensors rather than replacing them. Also, each milestone being accompanied by a clearly articulated value or RoI can help convince stakeholders regarding the investment need and help mitigate this challenge.

- **Societal challenges**: There is a perceived threat that intelligent automation enabled by Industry 4.0 is expected to make some manual jobs redundant. Another factor that risks the implementation of Industry 4.0 projects is resistance to adopting new ways of working, especially from experienced workers.

This section lists a few challenges that can impact the adoption of smart manufacturing; however, the benefits of adoption far outweigh the potential challenges. One of the main benefits is the automation of manual and routine manufacturing activities/jobs. A good example of eliminating manual steps in the manufacturing process is automatically quality-checking finished goods (the separation of faulty and non-faulty parts) by leveraging video analytics, which is the topic of our subsequent section.

Automatic inspection of finished goods or parts

One interesting use case in the discrete manufacturing space is the automated inspection of finished goods or parts. Finished goods are inspected by analyzing video feeds from digital cameras; by deploying the required machine learning (visual inspection) models on the **Device Gateway (DG)**, these models output a binary value indicating whether a part is defective or not.

> **Important note**
>
> The term *sensing device* is not limited to video feeds from digital cameras; alternate sensing technologies (such as infrared cameras and x-ray cameras) may very well be used to support specific operating conditions and use case requirements.

If a defective part is detected, a robotic arm will take that part from the conveyor belt and put it in a waste bin. These models are trained at a central server and refined models are pushed to a gateway from the central server. If the model is not able to identify the part with the required level of confidence (e.g., a new part being introduced), then those images are sent to a central server to further refine the model. In this way, there is a cycle of model refinement and deployment.

The use case along with the applicable patterns are shown in the following diagram:

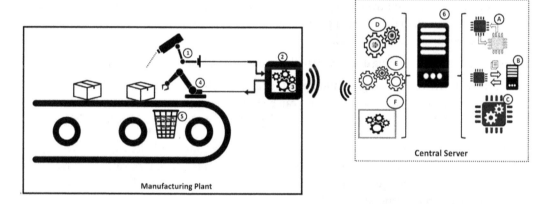

Figure 7.8 – Realization of the automatic inspection use case by leveraging IoT patterns

Let's look at this in further detail:

1. **Video camera**: A video camera would capture the images from the conveyor belt and this feed is sent to the DG. Communication between the video camera and DG would either be done using Wi-Fi or Ethernet.

2. **Device Gateway**: The DG receives the video stream from the video camera and identifies a part as *good* or *defective* based on the ML model that is deployed. If the model indicates the part is as fitting the specification, then that part is allowed to move further into the packaging

stage; otherwise, the DG instructs a robotic arm to pick up the defective piece and throw it into a waste bin. If the model is not able to identify a part with the required certainty, then that image is sent to the central server to be included in the training set.

This approach of part identification being done locally at the gateway has the advantages of bandwidth conservation (as not all the images are sent to the central server for analysis), as well as improved latency (avoiding the roundtrip to the central server for image processing/analysis and then sending the analysis back to the DG). To realize this use case effectively, analysis must be done in near real time so that instructions to the robotic arm can be sent on a timely basis. This also eliminates the possibility of the use case being rendered ineffective due to losing connectivity with the central server.

Although, in the preceding diagram, only one DG is shown, typically multiple DGs will be connected to the central server at a time and one DG would monitor multiple assembly lines/conveyor belts at the same time.

3. **Local Rule Engine** (**LRE**): An LRE will invoke the image recognition model for each of the incoming images supplied by the camera. Based on the output of the model, the LRE will issue commands to the robotic arm to pick the defective part up from the conveyor belt and throw that into the waste bin.

4. **Robotic arm**: The robotic arm receives specific commands from the DG/LRE to either pick the defective part up off the conveyor belt and throw it into the waste bin or do nothing and let the part pass on to the packaging stage. Generally, instructions from the DG to the robotic arm are sent using a PLC protocol.

5. **Waste bin**: This is for storing defective parts that shouldn't be allowed to pass to the packaging stage.

6. **Central server**: This is the aggregation point for all the DGs. This can be hosted on the public cloud (as a cloud-native application) or it can be hosted in an on-premises data center as well. There are additional IoT patterns that can be deployed on the central server, with details as follows:

    A. **Digital twin**: This will store the state of each of the conveyor belts with respect to its operational status (*in operation*, *halted*, *under maintenance*, etc.). It can also store metrics such as the percentage of defective parts per unit of time. This would also temporarily store the image-processing ML model before it is pushed to the DG.

    B. **File upload**: This will be used for sending the captured images (not recognizable by the current image recognition model). The same pattern will also be used when pushing the updated ML model from the central server to the DGs.

    C. **Device management**: This pattern helps manage numerous gateways that are connected to the central server along with establishing a hierarchical structure for easy manageability – for example, city > manufacturing plant > manufacturing area > assembly line/conveyor belt. Similarly, it can store other metadata related to the DG (e.g., the firmware version, certification rotation date, etc.).

D. **AI/ML integration**: This pattern is needed to create an ML model used for classifying a part on the conveyor belt as *good* or *defective*. As parts on the conveyor belt can come in multiple orientations, the ML model should be able to sift through all such images and still be able to classify them. As mentioned earlier, the trained model will be deployed on the DG for faster recognition and this model will be continuously refined based on the images that were not recognized by the current model. In other words, there is a circular interplay between the output generated by this pattern and the ML model that is deployed on the DG. The existing model is pushed to the DG and the DG sends a new set of images to the central server for model refinement. Once the model is refined, it is again sent to the DG.

E. **Enterprise system integration**: Enterprise system integration would be required to automatically order fresh raw material if the existing inventory goes below a defined threshold. For this, the number of finished goods as well as the count of defective pieces needs to be continuously monitored.

F. **Global Rule Engine (GRE)**: As mentioned in the previous point, a GRE will monitor events such as a low inventory of raw materials, the count of finished goods going beyond a defined threshold, and so on. Based on these events, the GRE will initiate relevant actions such as triggering supplier enterprise systems to order fresh raw material (in case of low inventory) or sending notification to the plant supervisor (in case of a low output of finished goods).

This section provided details regarding how IoT different patterns can be used to implement important use cases in the smart manufacturing domain and also brings us to the end of the chapter.

Summary

This chapter detailed the reasons why IoT is expected to play a vital role in making the manufacturing industry *smarter*. We started by defining a few concepts that are required for understanding the domain; outlined the key characteristics of five industrial revolutions; discussed the benefits expected from smart manufacturing; and, finally, covered how IoT patterns can be used to realize automatic inspection.

In this chapter, we highlighted the fact that in the future, manufacturing companies will act more like software companies by following agile manufacturing methodologies and data-driven decision-making. All this entails more collaboration and necessitates a big cultural change. Therefore, unless manufacturing companies actively invest in training employees, as well as address the expected cultural nuances, the aspiration of being a truly smart manufacturing enterprise will fail. An example of cultural change is to move from experience-based decisions to decisions based on hard data. Another cultural change is to empower employees to think creatively/innovatively about continuous improvement rather than merely following instructions. Supervisors are also expected to have more tolerance for genuine mistakes. It is expected that in the factories of the future, the proportion of software will gradually increase. In fact, the machines of the future will be very similar to the current smartphones on which the software is continuously updated – bringing feature updates and defect fixes.

In the next chapter, we will cover the agriculture domain and how IoT provides solutions to problems that the agriculture domain has struggled with for a long time.

8

Pattern Implementation in the Agriculture Domain

Traditionally, farmers spent most of their time observing plant and/or soil conditions and guesstimating the weather conditions before acting – for example, performing irrigation and applying fertilizers. The application of tools and techniques for **smart agriculture** helps remove this burden from farmers' shoulders, as decision-making can be performed by analytics engines.

Smart agriculture helps to reduce environmental impact through efficient irrigation and the optimal usage of fertilizer/pesticides. It helps to increase yield by obtaining accurate information about the soil and environmental conditions and then providing information regarding the optimum quantity of input (fertilizers, pesticides, water, etc.) and conditions. The problem with traditional farming is that farmers follow the same procedures regarding sowing, nourishing, irrigation, and harvesting without considering differences that exist in different areas of a field, resulting in unpredictability in farm yield and quality and resource wastage.

Agriculture has a unique set of challenges that are different from those in other domains such as manufacturing and retail, which underlines the need to understand this domain and how IoT and related technologies can be used to effectively mitigate those challenges. Some of these challenges that are agriculture-specific are as follows:

- High dependence on climatic conditions and an inability to determine/predict these conditions with the required accuracy.

- Farms are generally spread out over large areas and require a greater number of sensing and actuation mechanisms, which necessitates a high initial cost. Limited perimeter security also increases the chances of theft.

- Farm produce needs to be consumed within a short period of time and requires purpose-built storage and transportation facilities.

- Farming lands are normally located far from city centers and have limited connectivity

- The perception of agriculture being a relatively less glamorous domain attracts less technical talent, and correspondingly, less automation/digitalization is visible in agriculture compared to other domains, such as industry.

- A very small margin of error is acceptable in the farming process – a wrong decision can wreak havoc with a complete season's output. Also, there is no possibility of product repair/recall as there is with retail or industrial products.

Like other domains covered in this book, IoT works alongside other related technologies to provide interesting agriculture-related use cases.

Now, let's dive deeper into some key concepts in this domain.

An overview of smart agriculture

This section covers the key terms/definitions that are used in the *smart agriculture* domain. Understanding these terms is crucial to design and develop smart agriculture-related solutions.

Key terms/definitions

In this section, we will discuss some key technologies associated with smart agriculture:

- **Artificial Intelligence (AI)**: Accurate decision-making is the cornerstone of successful farming, and AI effectively complements/supplements a farmer's ability to make sound judgments. AI can be used in all stages of crop cultivation – for example, if a crop is infected with a disease, AI provides recommendations to reduce crop wastage. AI is also used to estimate optimal farm input, forecast demand, predict price/yield, identify crops that fit the soil/environmental conditions, and determine the right time to harvest.

- **Automation**: Automation can be used to start/stop the watering of crops (often remotely), either automatically or based on a predefined schedule. In **close loop automation**, the soil moisture level is sensed in near real time, and watering is stopped once the moisture level reaches a predefined saturation level. In more advanced systems, the weather forecast is taken as one of the inputs when deciding on the watering needs. If the weather forecast predicts rain, the watering of crops can be deferred. Similarly, by using pressure sensors alongside water pipes, water leakage/pipe rupture can be detected promptly to avoid water wastage.

- **Energy harvesting**: The power supply is not consistent in most farms; hence, there is a need to source the energy for operating sensors and so on from renewable sources, such as solar, wind, and hydroelectric power. Sensors and other IoT hardware can operate using battery power, but they bring the complexity of bulky equipment and the need for continuous replacement.

- **Blockchain**: Blockchain is required in an agriculture context to establish transparency and traceability in the *farm-to-fork* supply chain (the sourcing of farm input as well as the sale of farm output). Other aspects covered by blockchain include determining the fair pricing of farm output, the quick disbursal of crop insurance claims, and quick and easy credit. Blockchain can also secure field data from manipulation while in transit to the central server and while stored locally on the DG.

- **Robotics**: The vision of incorporating robotics in farm operations is to free the farmer from repetitive, monotonous, and mundane tasks. For example, an autonomous tractor can perform most farm activities, such as tilling, spraying, sowing, and harvesting. Robotics is also expected to increase the yield quality and quantity, as the operations are free from manual oversights/errors.

- **Big data analytics**: From a smart agriculture standpoint, this technology provides diverse benefits – for example, big data analytics is used for obstacle avoidance in autonomous vehicles, to predict yield quality and quantity and target prices, to suggest an optimal level of farm input, and to determine real-time operational decisions. In fact, it plays a vital role throughout the crop life cycle.

- **Aerial imagery**: Aerial imagery helps large-farm owners to monitor the state of farms (crop health, pest infection, irrigation problems, frost damage, and so on) in real time or to determine the changes in the farm state over a period. Aerial imagery can be acquired using satellites, drones, **Unmanned Aerial Vehicles (UAVs)**, or small aircraft, and these are typically equipped with cameras that have capabilities such as **Red, Green, and Blue (RGB)** and **Infrared (IR)** detection, **Light Detection and Ranging (LiDAR)**, and **Radio Detection and Ranging (RADAR)**. For example, using the **Normalized Difference Vegetation Index (NDVI)** is one such technique that uses a normal optical (RGB) camera along with an IR camera to determine crop health by calculating the amount of visible and IR light reflected by the crops, with a higher NVDI indicating healthy crops and a lower NVDI indicating unhealthy (pale yellow or brown color) crops.

- **Drones**: Drones are used in relatively large farms to assess crop health and pest or insect infestation, perform an overall field assessment, undertake planting (seeding), spray pesticides, and so on. Some drones are equipped with thermal, optical (RGB), or multispectral cameras to determine crop health/growth, over/under-watering, as well as expected farm output. The combination of cameras and a sprayer on a drone, along with strong analytics, enables you to selectively target areas of a farm that require pesticide spray, similar to a manual spray. Input captured by drones is fed into an analytics engine, enabling a farmer to have a more accurate prediction of the quality and quantity of the farm output and plan for downstream activities, such as optimum sale price/market selection and storage requirements.

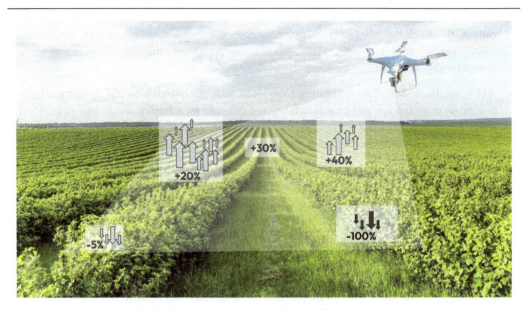

Figure 8.1 – Aerial imagery captured by drones

- **Remote sensing**: Vegetation indices calculated by using multispectral cameras help to determine the onset of crop disease promptly and indicate the correct time for crop harvesting. For example, if a vegetation map of a field shows that leaf color is changing from green to yellow, this indicates that nutrients have been passed from leaves to fruit, and the fruit is ripe for harvesting. Different types of optical, electrochemical, and mechanical sensors are used to determine soil and crop conditions. Optical sensors interpret data based on crop or soil pigmentation, and electrochemical sensors help to determine soil's electrical characteristics by measuring the concentration of elements such as potassium and phosphorus. In addition to measuring the direct soil moisture, there has been a recent trend to measure soil water tension (calculated as the water pressure of the soil), which gives a more accurate view of the water requirements, as the readings are not impacted by the soil type (i.e., factors such as soil texture, soil salinity, and organic composition).

- **Agronomic data**: Raw data accumulated from a farm and farm-related activities is referred to as agronomic data, examples of which include soil condition data, yield data, pesticide and fertilizer consumption data, and NVDI data. In an IoT context, this data can either be obtained directly using sensors or derived from systems that traditionally maintain this data (refer to the *Enterprise system integration* pattern in *Chapter 2*).

- **Geospatial analytics**: Geospatial analytics refers to the collection, transformation, aggregation, and visualization of imagery and locational data (**Global Positioning System** (**GPS**) coordinates) to determine insights, correlations (the impact of applying fertilizers in different parts of a farm or across farms), and trends (historical shifts as well as predictions) based on farming parameters. Geospatial data can be gathered from aerial imagery or connected farm equipment (tractors, combine harvesters, etc.) and can constitute both structured as well as unstructured data. Geospatial visualizations include overlaying current (as well as historical and predicted) sensor data over farm images.

- **Electronic Field Record (EFR)**: EFR refers to a data schema to store agronomic data. It describes the various entities (field and weather conditions, crop states, soil conditions including nutrient content and moisture level, and aerial imagery obtained from multiple sources) and the relationships between them. EFR is processed data (compared to raw data represented by agronomic data) and is used as input for analytics.

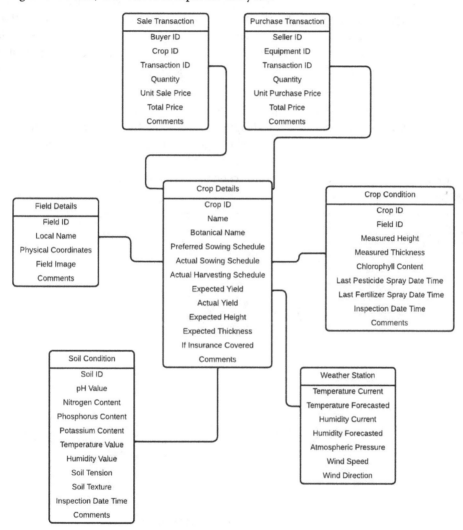

Figure 8.2 – A sample EFR

- **Hydroponic system**: A hydroponic system of cultivation involves growing crops with water as a key ingredient but without using soil. As oxygen is normally supplied via soil, an alternate mechanism of supplying oxygen is used – for example, the usage of air stones and air pumps is quite common. Similarly, support given to plant roots by soil needs to be provided by alternative materials, such as vermiculite, perlite, gravel, peat moss, coconut fiber, and rockwool. Nutrients that are naturally available in the soil such as magnesium, phosphorus, and calcium need to be artificially injected into the water stream.

A hydroponic system allows farmers to grow plants all through the year and at any place, and the output is generally more nutritious. Since the crops are grown in controlled conditions, they mature faster; typically, they grow faster by approximately 25%. They use less water due to less evaporation, and water consumption is reduced by 90–95%. Common crops that are grown using this mechanism include lettuce, tomatoes, peppers, cucumbers, strawberries, peppers, and cannabis.

IoT sensors monitor critical inputs required for the effective functioning of hydroponic systems such as water level, pH, temperature, and lighting conditions. These inputs are continuously analyzed to determine whether there is a need to regulate supply by sending instructions to IoT actuators. After analysis of the data, alerts/notifications can be sent to farmers if a deficiency (or oversupply) of inputs has a negative impact on the overall yield. This analysis is carried out by the LRE deployed on DG.

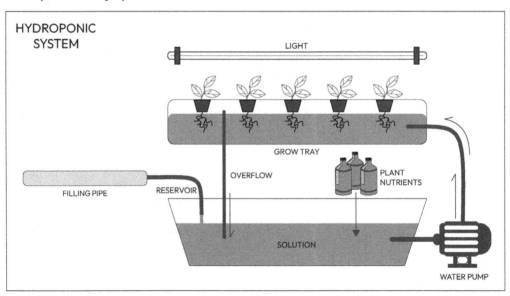

Figure 8.3 – Mechanics of a hydroponic system

- **Smart greenhouse**: A smart greenhouse creates an isolated and self-regulating environment that is customized as per crop requirements. Furthermore, crops no longer are at the mercy of weather/climatic conditions (wind, hailstorm, extreme temperatures, and ultraviolet radiation) and are protected from attacks from pests, locusts, and so on. It is a well-known fact that although some level of moisture/humidity is required for plant growth, excess humidity can adversely impact plant health as it encourages the growth of fungi and causes bacterial infections. Similarly, extreme temperature fluctuations can severely hamper overall production.

Smart greenhouses operate more in a *closed-loop* fashion, where data from sensors is continuously analyzed and the environment is regulated (by controlling spraying, irrigation, lighting, temperature, humidity, soil nutrients, etc.). This controlled environment helps save precious energy and is cost-effective in the long run. Typically, greenhouses are irrigated using drip irrigation (refer to the next bullet point), which also drastically reduces water consumption. Some of the more advanced greenhouses can open and close the enclosure based on environmental conditions (e.g., wind and rain).

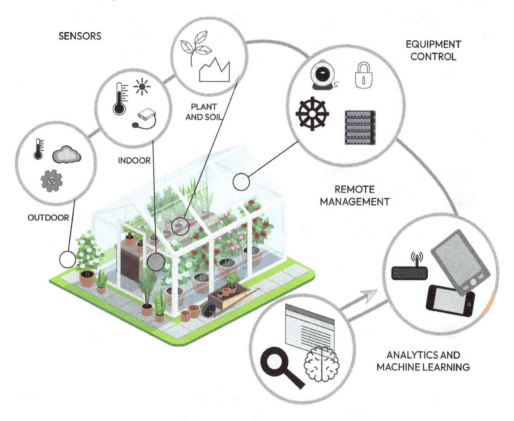

Figure 8.4 – The key components of a smart greenhouse

- **Micro irrigation/drip irrigation**: Micro irrigation (or drip irrigation) involves supplying water slowly using mechanisms such as mini-streams and droplets. This optimizes the usage of water (water reduction in the range of 45–70 % is observed) compared to traditional methods of irrigation (flooding), as water is applied to targeted areas, eliminating water runoff. There are some additional benefits of using micro irrigation, including the following:

- It reduces weed propagation (weeds are unwanted plants that grow alongside crops and compete for water and nutrients with the main crops). Micro irrigation helps to reduce/ eliminate the propagation of weeds, as water is applied directly to the main crops' roots, thus limiting water supply to weeds.

- It reduces the need for fertilizers, herbicides, pesticides, and so on, as these are supplied in soluble form and applied to the target crops directly and in concentrated form.

IoT can help realize the benefits of micro irrigation by ensuring that the optimum water supply is released to the plants and also that plant growth is not impacted by a reduced supply of water and other nutrients.

Figure 8.5 – Micro irrigation/drip irrigation

- **Site-Specific Crop Management (SSCM)/precision agriculture/internet of agriculture:** SCCM or precision agriculture refers to the concept of using a scientific approach to observe, measure, visualize, and respond to conditions that play a role in crop growth. It helps to reduce the variability in farm output as well as the quality. The objective is to understand the reasons behind variability in the yield and/or quality of farm output and take action to reduce the observed variability. This contrasts with traditional methods where farming practices are applied in a wholesale manner without considering specific crop needs and environmental conditions (e.g., water, nutrient, and sunlight requirements). The moisture content of different segments of a farm can be different, and once the variability is determined, water can be supplied to relatively dry areas by using drip irrigation.

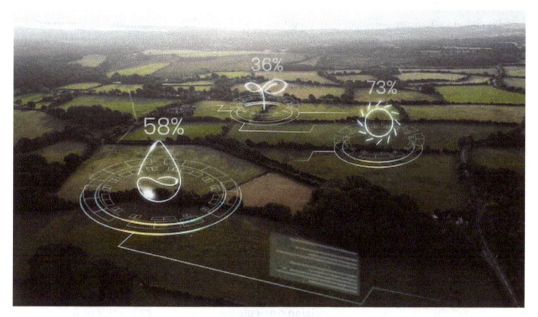

Figure 8.6 – Sensors monitoring plant and soil conditions

IoT plays a vital role in precision agriculture, with observation and measurement requirements fulfilled by deploying sensors (e.g., moisture sensors, pH sensors, and sensors that detect air speed/direction, etc.) and actuators that respond accordingly (e.g., automatic irrigation pumps). Data from the farm sensors is analyzed at a central server to generate recommendations related to planting, fertilization, irrigation, and harvesting. This *data-driven decision-making* approach contrasts with traditional farming, where decisions were primarily based on intuition or experience.

Sensor data is normally seen in correlation to a field's geospatial information, and as a result, GPS data needs to be captured alongside other variables. Accordingly, most of the stationary sensors used in SCCM have GPS capability built-in, or farm vehicles (tractors, combine harvesters) equipped with GPS capability can act as a **sensor hub**, where field data is obtained along with location data. A sensor hub helps to understand the farm's condition from multiple perspectives – for example, identifying areas with a high concentration of pests and weeds, and the overuse of fertilizers.

Along with IoT, the maturity of AI/ML and, specifically, the increased accuracy of predictions has played a vital role in the acceptance of precision agriculture by the farming community. As mentioned previously, the accuracy of predictions (e.g., expected weather conditions) plays a vital role in ensuring a higher yield, the quality of farm output, and a smaller environmental impact, as well as optimizing the usage of farm inputs (e.g., fertilizers and water).

Precision agriculture contrasts with traditional farming, where inputs were applied in fixed quantity/frequency and without considering the differences that exist in different parts of the same field. Similarly, the accurate estimation of a farm yield, along with the supply chain conditions, can help a farmer set an optimum price and estimate the amount of storage space required.

In fact, precision agriculture can be applied to all the life cycle stages of crop production, as shown in the following figure:

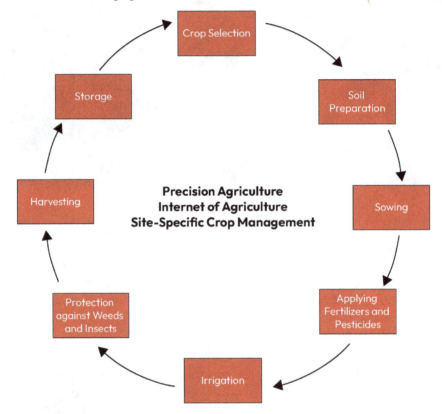

Figure 8.7 – Precision agriculture positively impacting different agricultural stages

- **Variable Rate Technology (VRT)**: VRT refers to the process of varying or adjusting farm inputs (e.g., fertilizers, water, and herbicides) based on the actual requirements of a field. **Variable Rate Application (VRA)**, **Variable Rate Irrigation (VRI)**, and **Variable Rate Seeding (VRS)** specifically focus on different types of inputs. As an example, VRS drastically improves a yield by minimizing seed wastage. VRS enabled by smart seeding equipment controls the depth and spacing of seeds to match different soil conditions.

- **Vegetation Index (VI)**: VIs are used to monitor crop health using multispectral techniques by measuring the amount of reflected/emitted electromagnetic radiation from crops. These indices help to understand the spatial and temporal variances in crop health (by measuring attributes such as chlorophyll content, soil moisture, etc.). Different vegetation indices are available and selected based on factors such as the crop type, the crops' development stages, crop density, field topology, or a specific plant attribute that needs to be measured. Some of the common vegetation indexes are NDVI (the most used vegetation index, as it can be used throughout the crop's development life cycle), the **Red-Edge Chlorophyll Vegetation Index (RECI)**, the **Normalized Difference Red Edge Vegetation Index (NDRE)**, and the **Modified Soil-Adjusted Vegetation Index (MSAVI)**. Calculation of these indices, as well as recommendations to consistently improve indices, are important applications of IoT in the agricultural domain.

As we are now aware of the key terms used in the smart agriculture domain, we are better equipped to understand how to implement these use cases in this domain. But before that, it will help to understand the reasons why it is currently gaining traction, which are described in the next section.

Factors influencing greater adoption of smart agriculture technologies

There has been increased adoption of smart agriculture practices recently. Let's look at some of the factors responsible for this transition from *traditional* agriculture:

- **High demand due to increased population and a change in dietary preferences**: As per the United Nations report of 2019, by the year 2050, farm output must increase by at least 70% to meet the food needs of the increased population. This increase in farm production must be done using existing farmland only, since it is not environmentally or economically possible to bring more land under cultivation. Smart agriculture is poised to solve this challenge by increasing yield on the same amount of land and making agriculture processes more efficient. Also, governmental agencies may incentivize farmers (in the form of equipment subsidies, free technical know-how, etc.) to adopt smart agriculture practices.

- **Data as an additional revenue stream for farmers**: Each of the crop life cycle stages such as seeding and harvesting generates a huge amount of data that can be leveraged by the farmer to enhance future crops, and it can also be sold to companies that use it to determine the key factors that determine farm output. This is a win-win situation for both farmers and companies, as farmers typically have limited knowledge of how to organize and leverage this data. Smart agriculture companies can help farmers easily derive insights from data by providing visualization and/or analytics services. For example, these insights can help farmers to select the right crop to plant and decide how many seeds should be sown per unit area.

- **Sustainability considerations/continuous depletion of water tables**: Around 70% of the fresh water on the planet is used by agriculture, and 30% of greenhouse gases are emitted by farming operations. The adoption of smart agriculture practices will drastically cut down water consumption, as well as drastically reduce environmental pollution. Additional benefits include the usage of fewer fertilizers and pesticides, which, in addition to having a severe environmental impact, also affect human health and well-being. In fact, the use of excessive pesticides is directly linked to conditions such as cancer and neurological diseases (refer to `https://www.ncbi.nlm.nih.gov/pmc/articles/PMC2231435/`, `https://www.beyondpesticides.org/resources/pesticide-induced-diseases-database/cancer`).

After understanding the reasons for the adoption of smart agricultural practices, let us delve deeper into some use cases where the application of IoT can provide an immediate benefit.

Use cases of IoT in smart agriculture

IoT can be implemented in various ways in the smart agriculture domain. These include the following:

- **Soil/crop health monitoring**:

  - Measuring temperature, moisture, pH values, and lighting conditions in real time
  - Determining chemical (fertilizer) composition
  - Wind speed/direction sensing
  - Frost detection/avoidance
  - Pest control
  - Climate monitoring and forecasting
  - Comparing crop growth, leaf size, and crop pigmentation with similar crops and conditions
  - Preventing soil degradation

- **Enhancing agricultural yield**:

  - Predicting the optimum time to plant, irrigate, and harvest crops
  - Predictive maintenance of farm equipment
  - Reducing paperwork while making insurance claims (e.g., crop/livestock damage)
  - Farm asset monitoring
  - Automatic irrigation
  - Aligning irrigation cycles with predicted weather conditions
  - Reducing crop damage via a disease or adverse environmental condition by leveraging predictive analytics
  - Evaluating and neutralizing the effects of the previous season's cultivation

- **Optimizing production cost**:

  - Optimizing energy consumption

  - Smart logistics and warehousing to help reduce grain spoilage in warehouses and grain elevators

  - Supply chain visibility and traceability

  - Demand and price prediction

  - Helping with decision-making – which crops to cultivate?

  - Leakage detection and reducing water wastage

  - Smart greenhouses to enable predator-free crop cultivation in a controlled environment

  - Vertical farming for space optimization

  - Drone monitoring

- **Livestock management**:

  - Disease detection and containment

  - Geofencing and location tracking

  - Livestock feed and produce statistics

  - Livestock theft prevention

  - Monitoring feeding/grazing patterns

  - Preventing consumption of non-edible (plastic) materials

  - Remote veterinary support

  - An automated feed supply to reduce the possibility of over- and under-feeding

  - Animal activity tracking and analysis

  - Monitoring milk and egg production

To understand this better, let's discuss one specific case of the application of IoT in this domain.

Resolving agricultural challenges with a land consolidation platform

Developing countries face unique challenges in the agriculture sector compared to developed economies. Most of the issues stem from the fact that the average landholding is quite small compared to developed countries. For example, India is characterized by a large number of small farms, with the majority of the land dependent on natural rainfall for irrigation purposes. This situation is further compounded by the fact that more than half of the Indian population has farming as a main source of livelihood.

A relatively small profit with limited insurance options results in a major chunk of farmers being perennially in debt. Additionally, there is the issue of land fragmentation, as land gets passed from one generation to the next. All these factors result in (constantly) dwindling profits per unit of farmland.

With limited automation, farming demands hard labor. This has resulted in most farmers' children quitting farming and choosing other (more glamorous) professions. However, the main stumbling block in the path toward mechanization/digitization is the capital expenditure required to adopt automation and the other smart agriculture practices mentioned in this chapter. Farm machinery required for automation is costly to the average farmer, and only rich farmers can afford such expenses. As the affordability of *smart techniques* is beyond the reach of the average farmer, improving the yield and quality of farm output seems difficult, resulting in a vicious cycle.

Apart from affordability, additional factors that block the smooth adoption of smart agriculture techniques are listed as follows:

- Erratic/intermittent network connectivity
- Unavailability of smart farm equipment and sensors for remote sensing
- Limited availability of trained experts that can guide farmers in their digitalization journey

Now, let's look at how these challenges can be addressed.

Mitigating agricultural challenges

Mechanization, along with smart agriculture practices, can be made possible by leasing agencies or by government assistance. The mechanized equipment (e.g., drip irrigation systems or remote sensing equipment) can help even small farmers to reap the smart agriculture benefits. These governmental/non-governmental agencies can also provide required training to farmers to run this equipment, as this will help to realize holistic benefits such as sustainability, hunger eradication, and a consistent food supply.

Another practical mechanism to reverse the vicious cycle discussed earlier is to implement a platform where small farmers can trade their holdings in return for **land shares**, similar to how shares get traded on the stock market. Corporates and/or bigger farmers can consolidate these holdings and apply automation and/or precision agriculture to enhance the yield, simultaneously reducing the cultivating cost per unit of land by leveraging **economies of scale**. In other words, the platform would facilitate farmers to lease or sell their holdings to third parties who can combine these small holdings to create a larger tract. This platform would enable farmers and third parties (which can be corporates or even farm bodies) to trade their land holdings for a set of land shares. Each small farmer would then receive returns proportional to the land shares they hold. A high-level architectural view of such a platform is depicted in the following figure:

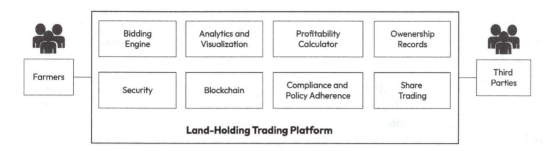

Figure 8.8 – A functional architecture of the land-holding platform

Although seemingly complex, this architecture can be realized by combining IoT patterns and generic software patterns. The key pattern that is relevant here is the DT pattern, as that would help a potential buyer gauge the fair market value of the land holding. Data in the **digital twin (DT)** would be provided by relevant sensors (e.g., temperature, humidity, pH, etc.). It would also help the buyer to understand the historical cost and profitability trends. Additionally, AI/ML would be required to make sense of the primitive data from sensors to estimate the potential yield from these holdings.

Additionally, a platform augmented by IoT technologies can help marginal farmers to get a fair price by growing crops that were hitherto considered not economically and/or environmentally feasible. Smart greenhouses allow farmers to cultivate crops that were grown in far-off places. Margins would improve, as there would be considerable savings on transportation costs.

The implementation of a land-holding platform using IoT architectural patterns and related technologies is illustrated in the following figure:

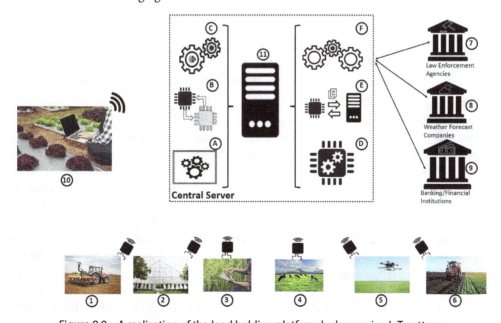

Figure 8.9 – A realization of the land holding platform by leveraging IoT patterns

The solution shown in the preceding figure is further detailed in the following list, where the requirements of a land-holding platform are integrated with the requirements of general smart agriculture use cases, such as livestock management, a smart greenhouse, and the operation of drones:

1. **Autonomous farm vehicles**: Automation is highly desirable in farming operations, as a typical farming season lasts for a few months and operations need to be completed in a minimal time frame. Autonomous vehicles are used to increase the operation time (24x7 operations rather than operations during the daytime only). Autonomous farm vehicles operate using other autonomous equipment – for example, they can detect obstacles using LiDAR, IR, and optical sensors attached to their periphery. Farm vehicles are either completely autonomous or semi-autonomous (where their path is remotely controlled using mobile devices such as smartphones or tablets – refer to the *Remote monitoring and control* bullet point). Most data processing and analysis (e.g., obstacle detection and route planning) is done locally on the local DG (a ruggedized gateway fitted inside the vehicle). Although completely autonomous operations (that can independently handle all the farm operations) are still decades away, some specific and repetitive operations are currently automated. The DGs communicate the vehicle diagnostics to a central server, which is used to predict maintenance schedules for the vehicles. Additionally, the AI/ML models that are developed or refined at the central server level are downloaded to the DG for accurate and efficient path navigation.

2. **Smart greenhouses**: Smart greenhouses can be considered self-contained IoT systems with sensors (e.g., light, CO_2, moisture/tension, pH, or temperature), DGs (for local processing), and actuators (sprinkler, air cooler, thermostat, lighting control, sprayer, or retractable roof/windows sliders). Of particular interest is the actuator, which opens a greenhouse's roof and windows if the environmental conditions are conducive to a plant's growth and retracts it if the environment is detrimental (e.g., an attack from a pest). In fact, a smart greenhouse is an ideal implementation of precision/smart agriculture, since all the factors related to plant growth can be comprehensively controlled. Smart greenhouses interact with a central server for data aggregation purposes and to enable monitoring and control from remote locations.

3. **Precision agriculture**: The use case for precision agriculture is quite like a smart greenhouse, with the difference being that conditions are relatively less controlled. Typically, farms are in remote areas with very limited connectivity; hence, the DG needs to buffer data if there is connectivity loss. These conditions also result in inconsistent or incorrect data being received at the central server. Hence, appropriate data messaging techniques (e.g., data filtering or checksum) are used before data can be analyzed/processed at the central server. Another challenge concerns supplying power to sensors and actuators, as remote farms are devoid of a continuous power supply. Hence, most sensors and actuators are powered by renewable energy sources (generally, solar power). This again brings added complexity, since energy consumption (by sensors or gateways) should be highly optimized. Also, the output of solar panels is impacted by ambient dust and other weather conditions such as rain and clouds. Generally, both connectivity and energy constraints are resolved by switching to lightweight communication technologies (e.g., UDP instead of TCP or technologies such as **Long Range (LoRa)**) and limiting the amount of

data (e.g., daily instead of hourly). Battery-powered field devices typically use aggressive power-saving techniques or algorithms (e.g., a device wake-up only if the current value is different from the prior sensor reading; otherwise, it is kept in sleep mode to save power and avoid redundant data). Another point worth noting is that as farms are open spaces, deployed IoT hardware should be ruggedized to withstand the vagaries of nature. Since farms are spread over large areas, automation (by way of autonomous farm vehicles, robotics, etc.) is widely employed.

4. **Livestock management**: Locations of the livestock are tracked using sensors such as GPS and RFID. These sensors help to create a **geofence**, whereby livestock is restricted to move in a particular geographic area. Alerts/notifications are issued to the owners if movement beyond the designated areas is detected. Input from activity-based sensors (e.g., cameras, microphones, accelerometers, or thermal sensors) along with feeding pattern information is used to determine/predict the health of farm animals and parturition conditions. AI/ML models for such predictions are developed/refined at the central server and deployed to local DGs.

5. **Drones**: Farm drones are used to observe crop and/or livestock conditions (e.g., over-fertilization, frost conditions, pests or diseases, irrigation issues, the calculation of VIs, or livestock movement) and aid in operations such as spraying and seeding. Most modern drones are equipped to monitor and take required action at the same time – for example, determining the areas infected with pests and selective application of a pesticide to reduce overuse, thereby optimizing costs. Route planning is normally done using customized software, and the route data is uploaded into drones at the launch site. Once drones are in flight, they can analyze the camera feed and send the analysis results (spatial coordinates of the farm areas with disease conditions) to a central server. One common approach is to downsample and send low-resolution images to the central server for real-time analysis, and higher-resolution images can be uploaded to the central server once the drone is back at the launch site where connectivity/bandwidth challenges are minimal. In general, real-time data (raw imagery or analysis results) is avoided to conserve the onboard battery.

6. **Smart farming equipment**: This refers to the general category of connected or smart farm equipment that is used for purposes such as automatic irrigation, the removal of previous crop residues, and the application of seeds, fertilizers, and pesticides. This equipment sends the operational and diagnostics data to a central server for fault detection and fault prediction purposes. As agriculture operations are seasonal and highly time-bound, any fault can adversely impact the yield and quality of the final produce, which makes it imperative that the equipment's faults are detected and rectified promptly. The operation of this equipment can be controlled remotely, as detailed in the *Remote monitoring and control* bullet point.

7. **External/third-party systems (law enforcement agencies)**: These are the external systems that are fundamental to implementing a land-holding system, as introduced in the prior section. A law enforcement agency's systems are required to administer and comply with the contract between farmers and third parties. **Enterprise integration patterns** (shown in *Figure 8.9* as part of the central server functionality) facilitate the interaction between central components and these external/third-party systems.

8. **Weather forecast companies**: Agriculture output relies heavily on accurately predicting/ estimating climatic conditions, and weather forecasts play an even more important role in smart agriculture, as they help optimize various activities. A simple example would be to postpone a scheduled/planned irrigation if the forecast data indicates rain. Similarly, fertilization would fail if rainfall occurred immediately afterward. Weather forecast data would be channeled by the enterprise integration pattern into a central server and used by patterns such as AI/ML integration and the global rule engine (defined later in this section).

9. **External/third-party systems (banking institution)**: Integration with banking systems is required to facilitate payments to farmers.

10. **Remote monitoring and control**: Farmers can use mobile devices to access insights generated by a central server. In scenarios where connectivity with the central server is erratic, mobile devices would connect directly with local DGs (e.g., to control the path traversed by drones or get a timely notification if an animal exits a defined geofence). The user interface design of mobile applications should be adapted to the needs and skill level of the farmers, and special consideration should be given to farmers who aren't very tech-savvy.

11. **Central server**: A central server can act as an aggregation point and host the following IoT patterns as well:

    A. **Global Rule Engine (GRE)**: GRE will help to coordinate the execution of defined actions if required conditions are met (e.g., start sprinklers if soil moisture is low, or send a notification to the farm owner if any farm animal shows signs of distress).

    B. **Digital twin**: The DT will store the current state and configuration related to each of the entities involved in a smart agriculture ecosystem (e.g., drones, farm equipment, farm animals, greenhouse, etc.). It would store this information along with the relationships of these entities in the real world – for example, a particular piece of farm equipment might be associated with one field, and that field can be part of larger farmland, and so on. As mentioned earlier in this chapter, connectivity is a challenge in remote farmlands, so the DT pattern becomes more important in this context, where the DT would store the configuration changes (done by the farmer, as mentioned in the *Remote monitoring and control* bullet point) and sync with the field devices once connectivity is reestablished.

    C. **AI/ML integration**: The prime purpose of this pattern is to develop and refine models that would be deployed locally to realize various **edge scenarios**. Autonomous farm vehicles and drones require these models for obstacle detection and to determine the need for preventive maintenance of farm equipment.

    D. **Device management**: This pattern helps to manage numerous gateways that are connected to a central server, along with establishing a hierarchical structure for easy manageability. Similarly, it can store other metadata related to the DG (e.g., a firmware version, a certification rotation date, etc.).

E. **File upload**: As connectivity (and sometimes even electricity) is limited in remote farmlands, farm data (e.g., sensor data from fields or livestock) is aggregated locally in the DG in the form of a physical file and pushed to a central server when the connectivity is available. DG keeps track of the part of the data file that is already sent to the central server to avoid the possibility of duplicate data being pushed to it.

F. **Enterprise system integration**: Implementation of smart agriculture requires integration with various third-party or external systems, as detailed in the preceding bullet points. This pattern will ensure that data with these systems is synchronized at the optimum time intervals (the frequency of synchronization will vary from one system to another) without putting extra load on the central server.

This section detailed the possible smart agriculture use cases, as well as the architectural patterns that are required to implement these use cases. One point worth noting is that integration with enterprise/third-party systems opens up possibilities of much richer use cases.

Summary

This chapter provided insights into how IoT plays a role in transforming the agriculture domain. Key challenges inherent in the agriculture sector were listed, and we discussed how IoT (along with other related technologies) can effectively tackle these challenges. Some of the key terms used in smart agriculture were explained. Lastly, a practical problem (smart farm holdings preventing the adoption of smart agriculture practices) prevalent in developing countries was highlighted and a practical solution was proposed.

This and the last few chapters focused on specific domains, and we illustrated how architectural patterns mentioned in the early chapters can be used effectively in these domains. The forthcoming chapters will cover some generic concepts, such as security, analytics, and edge computing. Accordingly, the next chapter will focus on the factors to be considered when selecting sensors and actuators for IoT use cases.

Part 3: Implementation Considerations

In this part, readers will understand the implementation challenges and table-stakes non-functional requirements, such as security, analytics, and the selection of sensors and actuators, that need to be considered while developing IoT solutions.

The intent is not to repeat what is already known about the non-functional requirements in question – the focus will be more on how these requirements need to be considered in an IoT context. For example, security is a well-documented subject with copious references available in the public domain; however, how security is applied in an IoT context (e.g., how security can be implemented for constrained devices) is rarely discussed.

This part comprises the following chapters:

- *Chapter 9, Sensor and Actuator Selection Guidelines*
- *Chapter 10, Analytics in the IoT Context*
- *Chapter 11, Security in the IoT Context*

9
Sensor and Actuator Selection Guidelines

Sensors and **actuators** constitute a critical part of any IoT system. As was mentioned at the start of this book, sensors are akin to human body parts of perception (eyes, ears, touch, and so on), whereas actuators can be compared to human parts of action (hands and legs). Through sensors and actuators, the IoT system interacts with the physical realm, and it is the ability of the IoT system to effectively blend physical (data acquisition and action) and virtual (processing) worlds that make IoT systems so unique and powerful.

In this chapter, we will understand the different types of sensors and actuators (the intent here is not to provide an exhaustive list of sensors and actuators but a representative list). Elements from these lists can be mixed and matched to develop innovative and interesting use cases. As we will understand in the chapter, these lists can be used in two ways:

- Given a problem statement, which sensors and actuators are best suited for solving that problem?
- Looking at the list of sensors and actuators, we can think of real-world problems that can be solved by a combination of sensors and/or actuators. Used in this manner, the list serves to demonstrate what's possible (the *art of the possible*).

This chapter provides guidelines/recommendations for selecting the right set of sensors and actuators. Toward the end of the chapter, we cover a brief introduction to the topic of **wireless sensor and actuator networks** (**WSANs**) as any non-trivial IoT use case would involve a network (or web) of sensors and actuators rather than individual sensors/actuators connected to the **Device Gateway** (**DG**). The section also lists a few techniques of how power consumption can be optimized when sensors and actuators are deployed as nodes of IoT networks.

Intricate/internal working of different sensors/actuators are deliberately omitted in this chapter for three reasons:

- IoT solution architects are expected to leverage available sensors/actuators in the market rather than design/develop them from scratch
- Knowledge of the internal functioning of the sensor/actuator doesn't help much in determining their actual applicability in solving real-world problems/challenges – it is more important to understand the purpose of these devices and the conditions under which they can be used

- Underlying technologies of sensors/actuators are continuously changing/evolving as per the advancements in the fields of material sciences and instrumentation

The next section explains terms and concepts that are typically used in relation to sensors and actuators. A high-level understanding of these terms/concepts will help you to compare the available sensors/actuators based on your specific characteristics.

Key terms/definitions

In this section, we define some key concepts associated with sensors and actuators:

- **Accuracy**: Accuracy refers to the ability of sensors to provide a result as close to real value as possible.

- **Precision**: Precision refers to the capability of the sensor to give the same readings for the same measurement over time and under similar conditions. Although accuracy and precision seem like similar terms, they differ in the sense that accuracy refers to how close the reading reported by the sensor is to the actual value, whereas precision refers to the ability of the sensor to detect even small changes. The difference between accuracy and precision can be better understood by the following figure:

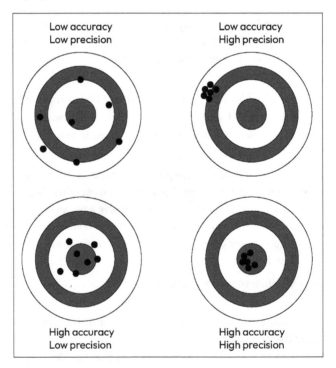

Figure 9.1 – Difference between accuracy and precision

- **Repeatability**: Repeatability indicates the ability of the sensor to sense a particular value and report it as the same value consistently (under similar environmental conditions) over a period – that is, repeatability refers to the consistency of the value reported by a particular sensor when the same value is read multiple times. Repeatability differs from accuracy, as shown in the following figure:

Low repeatability,
Low accuracy

High repeatability,
Low accuracy

High repeatability,
High accuracy

Figure 9.2 – Difference between accuracy and repeatability

More specifically, repeatability refers to the property of a single sensor and how it behaves over time, whereas precision is a relative property where one sensor's output is compared with other (similar) sensors. If fact, all three properties (accuracy, precision, and repeatability) are complementary characteristics and should be considered in conjunction for a more holistic sensor selection.

- **Range**: Range indicates the lowest and highest value of the physical quantity that a sensor can measure. Typically, range and accuracy have a negative correlation – a higher range results in less accuracy.

- **Resolution**: This refers to the smallest change the sensor can register.

- **Response time**: This is the time taken for the sensor's output to reach its final value.

- **Sensitivity**: Sensitivity is the ratio of the incremental change of the value of the sensor reading to the corresponding incremental change in the value being measured. In general, sensors with higher sensitivity should be preferred; however, very high sensitivity also results in readings getting impacted by ambient noise, for example.

- **Static/dynamic load**: This term refers to the load that an actuator can handle while static (static load) and while it is in motion (dynamic load).

- **Deadband**: This refers to the condition when there is no output from an actuator even though the actuator is provided with input excitation energy.

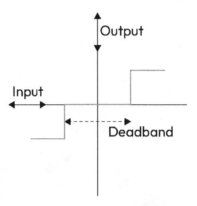

Figure 9.3 – Deadband illustration

- **Settling time**: Actuators employ mechanical parts/assemblies and as a result, they don't perform the action in one go and there is always an oscillation around the final/target position. Settling time refers to the time taken for this oscillation to stabilize.

- **Calibration**: Calibration refers to checking for the accuracy of sensors/actuators against a standard source and (if required) taking steps to rectify the deviation (drift) that typically sets in after continued use. Calibration is also required as a last stage in the production process as sensors/actuators produced using the same manufacturing process and equipment can differ in accuracy.

- **Linearity**: The graph between the value reported by the sensor and the actual value being measured should ideally be a straight line. Linearity refers to the extent of this variation from this ideal graph:

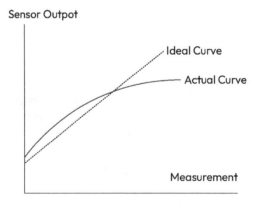

Figure 9.4 – Linearity resulting in differences between actual and observed sensor values

- **Drift**: Drift refers to the phenomenon of a sensor reporting a different value for the same reading if the readings are taken after a certain period. Similarly, an actuator will gradually stop reaching its exact target position (or sometimes move beyond the target position) due to continuous exposure to frictional forces or due to mechanical wear and tear.

When selecting a sensor, it is recommended that we perform a comprehensive analysis with respect to the device characteristics of the shortlisted devices as well as the environment in which the sensor would be used before placing the procurement order. The following section will help us perform this analysis.

Usage scenarios of sensors

Different types of **sensing technologies** are used in a complementary manner to implement complex use cases. As an example, autonomous vehicles leverage different technologies to accurately determine the environment (current location and approaching obstacles) and make sense of the available information. An autonomous vehicle best illustrates this as it uses diverse technologies to understand its environment in real time, as is illustrated in the following figure:

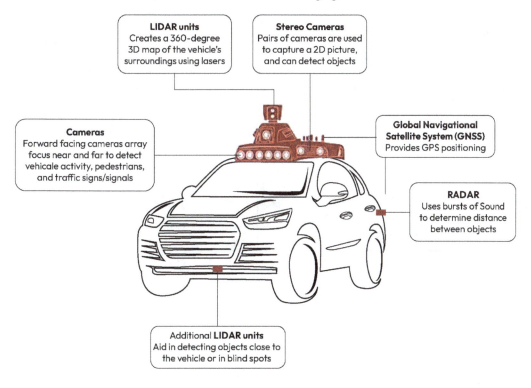

Figure 9.5 – Autonomous vehicle leveraging multiple technologies for understanding the environment

The following table compares the different sensing technologies used in autonomous vehicles and provides an example of how diverse sensing technologies can be used in a complementary way:

LiDAR	Video cameras	RADAR	GPS
Light Detection and Ranging; a source emits light that gets reflected from the object of interest. The reflection is then sensed by a LiDAR sensor to determine the time taken, which is then used for calculating the distance from the object of interest.	Digital video cameras – not very different from the ones found in mobile phones	Radio Detection and Ranging; similar to LiDAR, it uses reflection from objects/ surroundings to determine the distance with the difference that, here, radio waves are used instead of light	Global Positioning System; a mechanism for determining the current location that relies on determining the distance between the GPS receiver and a set of four or more satellites
+ 3D mapping + Ability to see in the dark and detect small objects	+ Ability to differentiate colors, character/text/sign recognition + Less expensive	+ Excellent range event at high speed + Works well in the dark and is unimpacted by ambient conditions such as smoke, dust, rain, and so on + Requires less processing power	+ Works in all weather conditions

LiDAR	Video cameras	RADAR	GPS
- Performance impacted by ambient conditions such as smoke, dust, rain, and so on - Expensive	- Inability to see in the dark and in conditions of smoke, dust, rain, and so on - Requires higher processing power	- Lacks precision and resolution and ability to detect small objects - Unable to differentiate colors, character/text and can't recognize sign	- Limited utility indoors and in closed spaces such as tunnels as technology requires direct line of sight with satellites - Less accurate and accuracy varies from region to region

Table 9.1 – Sensing techniques used in autonomous vehicles

Generalizing the example discussed, any IoT use case can be represented using the following equation:

$$IoT\ Use\ Case = (Sensor(Data\ Acquisition))^+ + Data\ Processing +$$
$$(Actuator(Data\ Processing\text{-}Based\ Action))^*$$

Here, $(Sensor)^+$ indicates one or more sensors (a representative list of sensors is provided in *Table 9.1*), *Data Processing* indicates processing locally (DG) or at the global (central server) level, and $(Actuator)^*$ indicates zero or more actuators (a representative list of actuators is provided in *Table 9.3*).

In other words, any IoT use case typically has *at least* one sensor, although it *may or may not* have an actuator.

For example, this equation, when adapted to the autonomous vehicle use case discussed earlier, would look like the following:

$$Autonomous/Connected\ Vehicle = (Sensors:\ Video\ Camera,\ GPS,\ LiDAR) + Data\ Processing +$$
$$(Actuators:\ Braking\ system,\ Steering,\ Horn,\ Front)$$

The following table lists the commonly used sensors in IoT solutions. The table also mentions the typical use case/relevant domain for each of the sensor types. As mentioned earlier, this is a representative list and not an exhaustive one:

Sensor Type	Applicability Details
Motion sensor	Detects unauthorized presence via detection motion; used in intruder detection systems
	Retail: Helps to design retail outlet layouts by determining the areas frequented by shoppers; automatic door opener
	Smart homes: Used for automatic door control, automated lighting control, automatic dispensing of water and soap
	Smart agriculture: Alerts farmers in case of any unauthorized movement (human or animal)
Alcohol sensor	Detects the concentration of alcohol (primarily ethanol) in the air
	Smart city: Used in breath analyzers to measure the alcohol content in vehicle drivers
	Industrial sector: Prevents accidents due to intoxication
Light sensor	Detects light intensity
	Industrial, Smart city: Used to adjust the brightness of the screen or other light sources as per the intensity of the ambient light
Smoke sensor	Smart city: Detects the possibility of fire/smoke
Rain sensor	Detects the amount of rainfall
	Smart agriculture: Helps to turn on sprinklers to avoid under/over watering
	Transportation: Used in high-end vehicles to automatically turn on wipers in the case of rain
Infrared sensor	Detects the thermal radiation emitted by objects/humans
	Smart city: Detects the presence of humans/animals and is part of night vision cameras
Chemical sensor	Industrial: Detects the chemical composition of objects, liquids, or gases
Flow sensor	Industrial: Detects movement of liquids and controls the flow of liquids in oil and gas industries
Gas sensor	Measures the concentration of different gases
	Industrial and Smart building: Detects the presence of hazardous gases and monitors indoor air quality
Gyroscope	Detects object orientation in three-dimensional space, used extensively for controlling the motion of drones and robots
	Athletics: Detects body posture and movement of athletes, which can be analyzed by coaches or guides to advise on bodily movements that would result in better performance

Sensor Type	Applicability Details
Water quality sensor	Ensures the quality of drinking water or water used for showers and swimming pools
	Smart city: Determines the quality and purity of the water, the concentration of undesired particles in the water, turbidity (suspended solid particles), and salinity
Accelerometer	Measures the acceleration of the object/vehicle
	Consumers: Used extensively in smartphones and for detecting free fall
Air quality sensor	Smart city: Detects the overall quality of air (number of harmful gases such as CO_2, sulfur dioxide, and so on) and particulate material
Magnetic sensor	Determines the presence and/or strength of the magnetic field
Pressure sensor	Used to measure gas pressure or atmospheric pressure
Humidity sensor	Measures atmospheric humidity
	Smart agriculture: Determines the need for irrigation
	Consumers: Used in consumer appliances such as refrigerators to ensure the freshness of food materials. Used in **heating, ventilation, and air conditioning (HVAC)** to ensure a comfortable environment
Noise sensor	Smart city: Measures ambient noise levels
	Smart building: Ensures the comfort level of the occupants and controls noise pollution
Proximity sensor	Helps to detect the presence of objects in the vicinity.
	Smart city: Determines the number of parking slots available and other obstacle detection/avoidance applications
	Home automation/assisted living: Detects unexpected movement. A blind stick is another application.
Temperature sensor	Reports current temperature and has applicability in almost all domains
	Smart agriculture: Ensures that crops are having optimal ambient conditions and alerts if the farm equipment is overheating
	Smart building: Used in HVAC to ensure that the temperature is maintained within a specified range
Level sensor	Detects level of liquids; typical applications include monitoring the level of water or water tank
	Smart city: Detects flooding conditions

Sensor Type	Applicability Details
Ultrasonic sensor	Used to measure the distance between two objects by using sound waves
	Used where ambient lighting conditions don't allow for image/optical sensors to be used
	Smart manufacturing: As the reflective property of sound varies depending on the intervening medium, the ultrasonic sensor is used effectively to detect empty parts on a conveyor belt (for example, to segregate filled and unfilled bottles).
Location sensor	Smart agriculture: Used to determine the altitude of the agricultural field along with latitude/longitude; geofencing for the livestock; aerial mapping of fields; and aids in the movement of autonomous farm vehicles
Optical/Image sensor	Detects various properties of light such as frequency, wavelength, intensity, and polarization
	Smart city: Detects water levels and helps in smart parking. Facial recognition determines a person's emotional and physical state (happy, sad, drowsy, and so on), as well as demographic characteristics (age and gender). Also used for gesture detection
	Smart agriculture: Determines crop health and helps in chlorophyll measurement and checking ripeness levels
	Retail/Industrial: Reads barcode labels
	Smart manufacturing: Detects malformed/defective parts on the assembly line
Mechanical sensor	Smart agriculture: Helps to measure the resistance offered by the soil by applying forces
Dielectric soil moisture	Smart agriculture: Determines the soil's water requirement
Electro-chemical sensor	Smart agriculture: Monitors the pH level of the soil, and helps to determine the level of minerals such as phosphorous, potassium, calcium, sodium, nitrogen, copper, and iron
Weather sensor	Smart agriculture: Acts as a sensor hub and measures aspects such as temperature, humidity, rainfall, atmospheric pressure, wind direction, and solar radiation
Salinity sensor	Smart agriculture: Used to measure the soil salinity
RFID (tag) sensor	Used for asset monitoring and asset tracking in multiple domains
Energy meter	Agriculture: Determines energy consumption for various farming operations
	Smart city: Energy can be measured at the aggregate level (complete household) or at the per-appliance level (such as HVAC energy consumption)

Sensor Type	Applicability Details
Anemometer	Smart agriculture: Measures wind speed and direction
Pluviometer	Smart agriculture: Used to measure rainfall
Volatile organic compounds (VOC) sensor	Smart building, consumer, smart city, and industries: Detects the presence of harmful organic compounds emitted by commonly found products such as paints, wood products, upholstery, stored chemicals, and so on
Thermal sensor	Retail, smart city, and industries: Determines the presence (position as well as count) of people/animals in a particular area by measuring body heat without violating privacy
Keypad	Consumer and industries: Starts/stops an operation and provides textual input

Table 9.2 – Representative list of sensors

As an exercise, you are encouraged to think of innovative use cases and to select the appropriate sensors and actuators (from *Tables 9.1* and *9.2*) for realizing those use cases.

Operation and usage scenarios of actuators

An **actuator** is a device that, based on some trigger mechanism, will make something move, rotate, oscillate, or initiate some operation. Actuators can perform actions on themselves (turn water sprinklers on/off, change temperature settings in a thermostat, and so on) or on the sensors (starting/stopping a sensor or moving a sensor from one location to another, for example). Generally, actuation requires more energy than sensing; accordingly, actuators are supplied with stronger batteries or are mains-powered. However, in a typical deployment, the count of sensors would far exceed the number of actuators.

For a completely automated and/or remote operation, sensors and actuator work in a complementary fashion to monitor and change the environmental state, as shown in the following diagram:

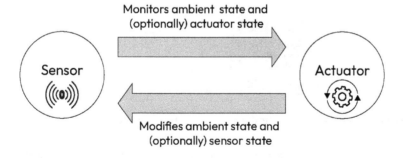

Figure 9.6 – Sensor and actuator working in a complementary fashion

Actuators can be broadly categorized into three categories based on the movement/action that is generated: **linear** (moves target in a straight line), **rotary** (generates a circular motion), and **oscillatory** (back and forth/pendulum type of movement). These different actions are depicted in the following figure:

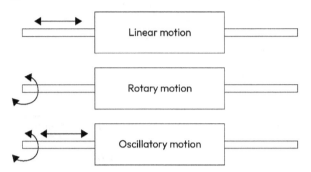

Figure 9.7 – Different types of actuator motions

Typically, an actuator converts electrical energy into force, motion, heat, voice, torque, and so on. To operate an actuator, two of the main inputs required are a **control signal** (trigger) and an **energy source**. An actuator can also be viewed as an energy convert/transformation agent where it converts electrical, mechanical, hydraulic, and pneumatic energy into mechanical energy (or mechanical movement), as depicted in the following figure:

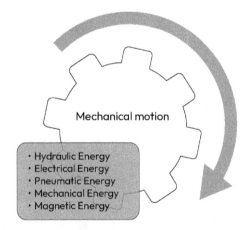

Figure 9.8 – Actuator converts different types of energy into mechanical energy

Each of these input energies has unique characteristics (trade-offs), which play a crucial role in deciding the right actuator. The pros and cons of each of these input energies are detailed in the following section.

Key characteristics of actuator types

This section lists different types of actuators that are available on the market. Each of the actuator types listed in this section has vastly different characteristics and understanding these characteristics is crucial for us to determine the suitability of these actuator types for a particular use case and/or operating context:

- **Electrical Actuators**: These convert electrical energy into mechanical energy. These electrical actuators can be further categorized depending upon the type of technology/mechanism used for generating required action (for example, **Brushless Direct Current Motor** (**BLDC**) and **Permanent Magnet Synchronous Motor** (**PMSM**) are two common types of electrical motors used for generating rotary motion). The pros of using such actuators are as follows:

  - Higher precision and better control

  - Require less maintenance and provides easy installation and operation

  - Less pollution

  - Less noise and highly efficient

  - No leakage risks

  However, the cons are as follows:

  - Can't work in harsh conditions or explosive environments

  - Expensive

  - Low speed

  - Can't support heavy loads

  - Occupy more space

  - Can't operate without an active power source

- **Hydraulic Actuators**: These convert hydraulic energy (fluid such as oil) to mechanical energy. Some of their benefits are as follows:

  - Durable and rugged – can work in harsh conditions

  - High power – can carry heavy loads

  - Provide high speed and precision

  - Can operate in harsh conditions and/or explosive environments

  - Can operate without an active power source

  - Self-lubricating

Some of the disadvantages are as follows:

- Prone to leakage and requires regular maintenance

- Limited acceleration

- Difficult to operate

- Pose a fire risk

- Expensive

- Noisy operation

- Can't work in cold/freezing environments

- **Pneumatic Actuators**: These convert pneumatic energy (vacuum or pressurized gas) into mechanical energy. Their advantageous characteristics include the following:

  - Durable and rugged, and can work in harsh conditions/explosive environments

  - Quick response

  - Lightweight, low cost, and require less maintenance

  - Produce large output force from relatively small input

  - Easy to install and have a long life

  - Can operate without an active power source

  - No leakage risks

The disadvantages include the following:

- Difficult to operate and require regular maintenance

- Less precise/accurate

- Noisy operation along with high vibration

- Can't support heavy loads

- **Magnetic/Thermal Actuators**: These convert thermal (temperature changes or magnetic energy) into mechanical energy. They are particularly useful due to the following:

  - Lightweight and compact

  - High power

  - Operate without an active power source

  - No leakage risks

However, they do have certain negative characteristics, which include the following:

- Slow response
- Less precise/accurate
- Can't support heavy loads

Now that we've learned about all the different types of actuators, let's look at some of their use cases.

Use cases for actuators

Similar to *Table 9.2*, which provided a list of sensors, the following table is the representative list of actuators along with applicable domains and use cases.

Actuator Use Case	Applicability Details
Ventilation systems	Smart agriculture: Ensure that crops (and livestock) are provided with favorable environmental conditions and allow fresh air into greenhouses
Fertilizers/pesticides/ seed spreaders	Smart agriculture: Useful in large farms for uniform application of fertilizers, pesticides, herbicides, and so on
Sprinklers	Home automation and smart building: Used to control fire by releasing water Smart agriculture: Irrigate agricultural crops in a manner that mimics actual rainfall
Automatic milking systems	Smart agriculture: Used for extracting milk from cattle
Automatic door openers	Smart city and retail: Automatically open the door when human presence is detected
Seat movements	Smart city: Seat adjustments in automobiles
LEDs/lights	Smart manufacturing: Indicate assembly line status
Buzzers/speakers	Smart manufacturing and smart city: Sound an alarm
Robotic arms	Smart city: Enable pan, tilt, and zoom functionality for surveillance cameras
Cooling fans	Consumer and industries: Provide cooling air to persons or to industrial machinery
IR blasters	Consumer: Used to send **infrared** (**IR**) commands from a remote control and simulate the remote control functionality Can be extended to control any appliance via gesture/voice commands
Relays	Consumer: An electrically operated switch used to trigger ON/OFF actions
Vibrations	Consumer: Provide vibrations in mobile and other electronic devices
Brake pedals	Transportation: Regulate vehicle speed by enabling the braking mechanism

Actuator Use Case	Applicability Details
Heaters	Consumer and industries: Used to heat target objects or supply heat to the environment

Table 9.3 – Representative list of actuators

To better understand the interplay of sensors, actuators, and data processing inherent in any IoT use case, let's delve deeper into the connected coffee vending machine use case in the next section.

Use case – connected coffee vending machine

The following figure shows the overall functionality of the connected coffee vending machine and the high-level interaction between the user and the coffee machine:

User can operate the connected coffee machine either remotely (via mobile devices such as smart watches and mobile phones) or directly from the coffee machine's UI. Similarly, users can receive machine responses either remotely or directly on the machine's human machine interface (HMI), which may consist of a screen or audio.

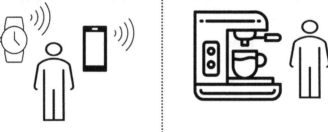

User settings and the information reported by the coffee machine include:

- Coffee type, ingredients, and so on
- Payment for the selected type
- Notifications (on the machine or mobile device):
    - Beverage completion status, such as "Drink complete"
    - Operational alerts, such as "Milk level low"
    - Diagnostic events, such as "Heating element behaving abnormally"

Figure 9.9 – Coffee vending machine operation and user interaction

For the coffee machine to provide the functionality specified in the preceding figure, various sensors and actuators need to be integrated and housed in assembly. In addition to the core data processing engine, diverse sensors and actuators would be required, which are shown in the following figure:

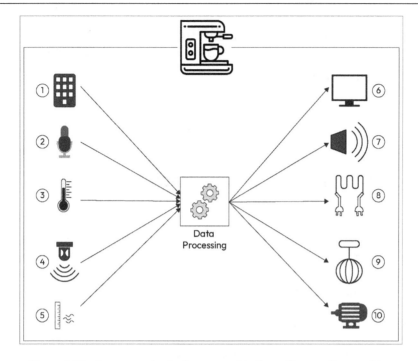

Figure 9.10 – Sensors and actuators present in the coffee vending machine

Let's look at the elements shown in the figure in greater detail:

1. **Keypad**: The keypad serves as a primary input mechanism for selecting the specific coffee types and quantity of ingredients such as milk and sugar. Also, users can schedule the brewing process to start at a later/predefined time. These inputs can be provided directly using the keypad/touchscreen provided on the coffee vending machine, or they can be specified using an application running on mobile devices (as depicted in *Figure 9.10*).

2. **Mic**: This is used for receiving voice commands/instructions from the user – as an alternative to the keypad.

3. **Temperature sensor**: This is used to measure the water temperature.

4. **Presence sensor**: This is used to determine the presence or absence of a cup to avoid coffee spillage.

5. **Water level sensor**: This ensures the right amount of water or milk is poured into the cup and helps to avoid conditions of overflow or underflow. Although not shown in the figure, similar sensors would be required to determine the level of other ingredients, such as milk, sugar, and coffee.

6. **Screen**: The results of the user selection as well as the progress of the brewing operation and notifications such as `sugar quantity running low` are shown on the display screen attached to the coffee machine. In the case of remote operation, similar information would be displayed using the mobile application UI and/or using push notifications.

7.   **Speaker**: Users can be updated on the status of the brewing process as an alternative to the screen.

8.   **Heating element/coil**: This is used to boil the water to the desired temperature.

9.   **Mixer**: This is required to mix/shake the ingredients.

10. **Motor**: This is used for dispensing the ingredients into a mixer container.

Now let's look at the important considerations that you should keep in mind while selecting a sensor or actuator.

Factors to be considered while selecting a sensor or actuator

Selecting a sensor and/or actuator is one of the most important design considerations as it involves balancing diverse (and often conflicting) requirements. As the number of sensors/actuators deployed in real/practical use cases is large, it is almost impossible to replace deployed field devices if they are found to be unsuitable later. Hence it is better to perform the required due diligence during the initial selection stage. Accordingly, this section provides guidance regarding the key factors that should be kept in mind while shortlisting the field devices:

- **Data and usage considerations**: This covers the type of data required and its purpose and it will cover aspects such as the type and size of the sample. Most of the applications perform the following collectively or in isolation:

  - Object recognition

  - Object presence/absence

  - Level/quantity monitoring

  - Distance ranging

  - Initiating movement either linear, rotary, or oscillatory

  Even in the same use case, attention should be given to the specific application area – for example, in smart agriculture, a sensor can be used to determine the level of water (depth of water table) or it can be used to determine the moisture at the soil surface. **Latency** is an important factor that should be considered – for example, fire or intrusion detection requires very low latency, whereas some applications such as moisture detection can work with higher latency.

  The same characteristic can be measured with different sensors (the weight of the farm animal can be measured using a weight sensor and can also be determined by running image recognition algorithms on the image captured by image/optical sensors). For actuators, the use case/problem domain would dictate the type of movement/action an actuator should trigger (linear, rotary, and so on).

- **Desired properties/attributes**: It is important to list the requirements related to the accuracy, precision, range, and other attributes required in a sensor to select a right fit sensor for the use case – for example, having more accuracy/precision than is required might not serve any purpose other than increasing the cost. Sensors monitoring health parameters would be expected to have much higher accuracy/precision than (say) sensors deployed in the manufacturing domain. Other important characteristics worth considering are the sensor's frequency response and the overall size/form factor of the sensor/actuator.

- **Operational environment/regulatory requirements**: The usage of sensors/actuators within a particular environment (ambient temperature/humidity, vibration/lighting conditions, and surrounding noise levels) governs factors such as enclosure design, mounting options, waterproofing, and so on. Environmental factors also govern the right **International Protection (IP) rating** – for example, IP55 (protection from dirt, dust, and water jets but can't sustain submerging in water) and IP67 (sensor/actuator protected from dirt and dust and can sustain being submerged in water up to one meter). For some of the domains (such as healthcare), sensors/actuators also need to comply with specific safety/regulatory standards such as **National Electrical Manufacturers Association (NEMA)**, **International Electrotechnical Commission (IEC)**, and so on.

- **Power requirements**: Whether the sensor or actuator would be operating from a mains power supply or battery operated or using an energy harvesting (such as solar energy) mechanism is an important consideration. Battery-powered sensors and actuators are easier to install but have the inherent issue of replacement after a few years. The operational environment also plays an important role in determining the right type of power source. For example, long-lasting batteries would be preferred for remote or harsh areas (such as underwater) to reduce the replacement hassles.

- **Connectivity requirements**: The connectivity technology (cellular, Wi-Fi, satellite, wired, and so on) used to transfer data from the sensor to the DG or directly to the central server is an important consideration. Connectivity technology selection depends on a number of factors, such as the number of nodes to be supported, the power or energy budget, and whether the processing is required to be done locally (DG) or remotely (central server). Another factor worth considering is whether the connectivity protocol follows an open standard or is a custom/vendor-specific protocol. In the case of the latter, vendor lock-in is an inherent risk.

- **Coverage considerations**: Some sensors require full coverage (such as image sensors or surveillance cameras), whereas others can make do with partial coverage (such as temperature or humidity monitoring in smart agriculture). In the case of partial coverage, care should be taken to place the sensor at a location that is a good representative of the general case. Also, the data retrieval frequency should be set so that adequate data is available for analysis.

- **Supplier/procurement considerations**: Some non-technical aspects such as supplier's lead time, type, and length of post-sales support (accuracy of sensors/actuators tend to drift and they require regular calibration), availability of alternate vendors providing similar devices, and commercial considerations (one-time versus recurring payment terms) also need to be considered. Some vendors operate in close ecosystems (data from/to the sensors/actuators flows not to the user's central server but to the vendor's central server from where data needs to be fetched), which not only results in inefficiency but also in security or data privacy issues. Additionally, in the case that both a sensor and an actuator need to be procured, preference should be given to the vendor who can provide both to reduce incompatibility issues.

  Another important point to consider is that the value reported by a sensor varies with different environmental conditions (such as ambient temperature, humidity, and noise levels) and hence it is crucial that these conditions closely mimic operating conditions specified or recommended by vendors.

- **Convenience/utility considerations**: This includes factors such as whether sensing or actuation information can be fed directly to the central server, or whether some manual intervention is required (download data from the sensor and then upload to the central server, for example). Similarly, sensors and actuators that can be configured or calibrated (smart sensors) remotely are generally preferred.

After understanding the factors that need to be considered while selecting sensors/actuators for implementing desired use cases, let us understand how they can be connected to each other through diverse network topologies.

Introducing wireless sensor and actuator networks

WSANs are generally controlled by a DG (also referred to as sinks or base stations). Mostly, WSANs are deployed in remote or obscure locations where they aren't mains powered, which necessitates the need to employ aggressive energy optimization techniques (inducing regular sleeping/waking cycles and minimizing the data communication) so that batteries can last for a longer duration (unless energy is harvested from the environment, such as solar/vibrational energy).

Another desired property from WSANs is the ability to add or remove nodes (sensors or actuators) without impacting the overall performance of the WSAN. To achieve this, nodes need to be designed intelligently whereby they can dynamically route data packets with any change in network topology. Typically, this is achieved by following a distributed architecture whereby each node acts independently and maintains the information about its nearest neighboring node. WSANs are typically deployed in the topologies that are shown in the following figure:

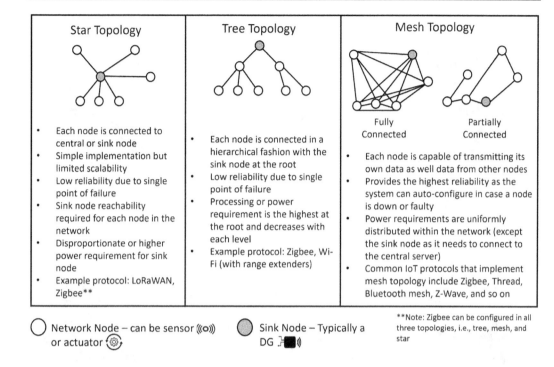

Star Topology

- Each node is connected to central or sink node
- Simple implementation but limited scalability
- Low reliability due to single point of failure
- Sink node reachability required for each node in the network
- Disproportionate or higher power requirement for sink node
- Example protocol: LoRaWAN, Zigbee**

Tree Topology

- Each node is connected in a hierarchical fashion with the sink node at the root
- Low reliability due to single point of failure
- Processing or power requirement is the highest at the root and decreases with each level
- Example protocol: Zigbee, Wi-Fi (with range extenders)

Mesh Topology

Fully Connected   Partially Connected

- Each node is capable of transmitting its own data as well data from other nodes
- Provides the highest reliability as the system can auto-configure in case a node is down or faulty
- Power requirements are uniformly distributed within the network (except the sink node as it needs to connect to the central server)
- Common IoT protocols that implement mesh topology include Zigbee, Thread, Bluetooth mesh, Z-Wave, and so on

Network Node – can be sensor ((o)) or actuator

Sink Node – Typically a DG

**Note: Zigbee can be configured in all three topologies, i.e., tree, mesh, and star

Figure 9.11 – Different types of sensor and actuator networks

As detailed in earlier chapters, the DG is primarily responsible for aggregating and analyzing data received from sensor nodes, providing connectivity to the central server (in some cases), making decisions, providing a controlling and monitoring interface to users or the external world, and implementing those decisions via actuators.

Another strategy used for optimizing power, especially in networks that are implemented in a large geographical area, is to add **mobility** to either sensor/actuator nodes or sink nodes. This is especially relevant in cases where these nodes can piggyback on some other entity. A DG placed on a tractor reading the data from the nearby sensors by forming an ad hoc network is an example of a dynamic sink node, whereas sensors placed on farm animals that get read when these animals return to the barn can be a good example of mobile sensor/actuator nodes. Adding mobility also solves another problem inherent in WSANs – non-uniform power required by different nodes in the network. **Node mobility** helps to determine the optimum coverage and power consumption by placing nodes at different physical locations and comparing the overall coverage/power consumption (a *hit-and-miss* approach). Sensor/actuator nodes as well as sink nodes can either be stationary or they can be mobile (as per use case requirements). The possible scenarios with one example application are shown in the following figure:

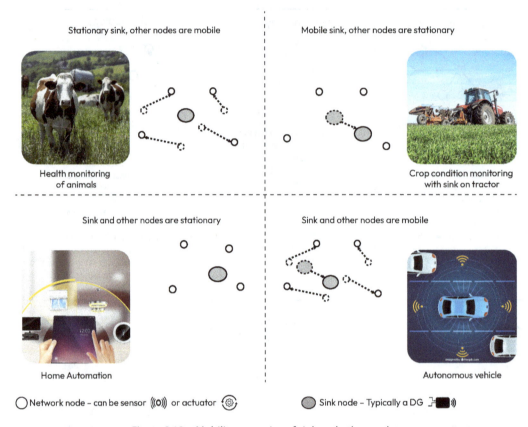

Figure 9.12 – Mobility scenarios of sink and other nodes

As can be seen in the preceding image, the energy requirements of nodes nearer to the sink are much higher than nodes that are far from the sink, as nodes near the sink are required to transfer more data.

One optimization strategy that is used by WSANs is to perform **filtering** and/or **data aggregation** at intermediate nodes (such as aggregation or filtering done at the sink node). For example, if neighboring nodes are sensing similar values, then instead of sending similar (and redundant data) to the sink, only filtered and aggregated data can be forwarded, reducing data traffic, and correspondingly, reducing power consumption. Another mechanism used is called **change of value** (**CoV**), where instead of *polling* for a network node value repeatedly, the sink node sends a subscription request to the network nodes. This allows the network node to send an updated value only if it changes by a specific amount (delta value, where the delta value is sent initially as part of the subscription request); all other changes (where change value < delta) are ignored, thereby conserving precious energy and bandwidth.

Data compression is another technique used to optimize data traffic within the network. Another technique that is used to reduce the traffic between the sensor and sink node is to **predict** and use future values rather than sending them over the air/wire. One such technique is illustrated in the following figure:

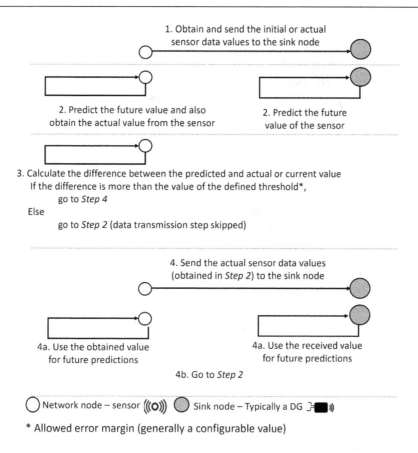

1. Obtain and send the initial or actual
sensor data values to the sink node

2. Predict the future value and also
obtain the actual value from the sensor

2. Predict the future
value of the sensor

3. Calculate the difference between the predicted and actual or current value
If the difference is more than the value of the defined threshold*,
    go to *Step 4*
Else
    go to *Step 2* (data transmission step skipped)

4. Send the actual sensor data values
(obtained in *Step 2*) to the sink node

4a. Use the obtained value
for future predictions

4a. Use the received value
for future predictions

4b. Go to *Step 2*

Network node – sensor ((o))   Sink node – Typically a DG

* Allowed error margin (generally a configurable value)

Figure 9.13 – Optimizing the data transfer by predicting future values

As can be seen from the preceding figure, WSAN design involves a trade-off between coverage, data accuracy, and energy consumption. Nodes other than sink nodes are constrained in terms of compute, connectivity, storage/memory, and power/energy availability. These constraints not only make designing an energy-efficient network challenging but also severely limits the ability to implement non-functional requirements such as security, as most of the security algorithms are resource intensive. As a result, nodes often transfer unencrypted data, directly exposing it to be exploited by bad actors. These challenges (specifically related to security) and their mitigations are detailed in *Chapter 11, Security in the IoT Context*.

WSANs are generally placed at desired locations and are expected to auto-configure once they are powered on. To conserve power, these nodes are designed to have a very short transmission range and are capable of single-hop communication (sending data to the nearest neighbor) only. However, after being configured as part of the network, these nodes can send data to any node (making an arbitrary number of hops). As mentioned earlier in this section, transmission normally terminates at the sink node.

Summary

This chapter covered the sensors and actuators that form crucial elements (similar to the eyes and ears) of any IoT network. A representative list of sensors/actuators was also provided as an aid that can help you while developing an IoT solution for novel use cases that can be implemented using a combination of the suggested sensors/actuators. Two specific examples (one related to a connected coffee vending machine and another to autonomous vehicles) were provided to illustrate how a diverse set of sensors and actuators are needed for implementing real-life IoT use cases. By now, you should be able to grasp key characteristics that need to be considered while selecting a sensor and/or actuators for the envisaged IoT use case.

You were also introduced to the different topologies in which sensors and actuators can be arranged (WSAN topologies) to serve diverse operating needs. Insights were shared regarding techniques or tactics used to optimize the data transfer in WSAN with an overall objective of reducing bandwidth and/or power requirements.

This chapter covered sensors and actuators that are primarily involved in data acquisition and act in the physical world, but that action can only be taken based on insights gleaned from the accumulated data. These insights are generated by running analytics on top of gathered data; this is what we will focus on in the next chapter.

10
Analytics in the IoT Context

In any non-trivial IoT use case, a huge volume of data is generated at a high speed. This high-volume data needs to be analyzed at similar speeds so that meaningful insights can be deduced, and the required actions can be triggered quickly. Most of the advancements in (generic) analytics can be applied directly to IoT use cases, but two key characteristics of **data ingestion** (that is, high volume and high frequency) necessitate that some special considerations are taken while reusing generic learnings/algorithms in the context of IoT. For example, IoT visualizations (dashboards) need to be displayed at reasonable granularity while not missing out on crucial/anomalous data points.

In addition to data volume and data velocity, IoT data is different as it can be a combination of **structured** (sensed values in time series format, such as temperature values captured at intervals of 1 second, and inventory data), **semi-structured** (operator comments), and **unstructured** data (video/image data). This chapter will start by covering the key terms and definitions that are relevant to IoT analytics and then cover IoT-specific nuances that you should be aware of while applying analytics.

Often, data that's captured by resource constraint devices/sensors is inconsistent and/or inaccurate (the sensor is running on a low battery and is reporting an inaccurate reading, for example), so this data needs adequate massaging before it can be used for analysis. This is why **data cleaning** is a major requirement in IoT use cases. A high volume and/or velocity of data ingestion necessitates that these cleaning activities are automated. Due to this, data cleanup activities are assigned to **artificial intelligence/machine learning** (**AI/ML**) algorithms. We will cover this aspect of ensuring **data quality** in reasonable detail.

Some use cases require data to be processed/analyzed locally rather than at a central server due to requirements such as low latency, data security/privacy, and more. Processing data near its point of ingestion rather than processing it at the central server is referred to as **edge analytics** and is something else we will cover in this chapter. Lastly, we will cover specific points that need to be considered while presenting analysis results to the user – that is, IoT **visualizations**.

Key terms/definitions

In this section, we'll look at some key concepts that are relevant to IoT analytics:

- **AI**: AI intends to replicate human intelligence by using systems that can learn from past decisions, predict future scenarios, and continuously improve decision-making capabilities. AI has special relevance in the IoT context as the data that needs to be processed is high in volume, velocity, and variety, as discussed in the *An overview of IoT* section in *Chapter 1* (the seven Vs of IoT data). This can't be processed by traditional computing systems that are strictly rule-based (if X happens, do Y) and can only serve a very narrow purpose. Complex decisions need to be made based on the values in the data stream that are beyond the capabilities of traditional computing/programming systems. *Chapter 12*, *Exploring Synergies with Emerging Technologies*, describes in detail how AI and IoT act as complementary technologies to solve real-world challenges such as the following:

  - Autonomous vehicles use a combination of IoT and AI to make real-time routing decisions

  - Data gathered by sensors in the manufacturing industry domain can be analyzed by AI algorithms to generate *process optimization* recommendations

  - In the retail domain, AI can analyze shopper movements to provide predictions such as checkout waiting times, footfall-to-sales ratios, and more

- **ML**: ML works on the principle that a computer program can autonomously improve its performance by learning from available data. This is different from procedural/rule-based programs (if X, then Y).

  IoT data corresponds to real-life events, which have inherent unpredictability (not all the events can be known beforehand), and traditional procedural/rule-based programs, which have limited applicability in the IoT context. ML can effectively fill this gap and differs from traditional analytics in the following ways:

  - ML can consistently learn and unlearn new trends/outcomes. It can be used to compensate for the limitations imposed by the constrained nature of field devices by enhancing their computing power, optimizing their power consumption, and more.

  - ML techniques can be used to provide important situational and/or behavioral contexts, which are required for more comprehensive decision-making and automated realization of those decisions. ML works by using the available dataset for model creation purposes; then, the trained model is applied to real data. Whereas traditional analytics builds a model based on past data and expert opinion, ML starts with an outcome or goals (such as energy conservation) and then works backward to determine factors that influence the outcomes and their relationships. In other words, it *learns* from the data factors that have more influence in achieving the defined goals.

- The performance of traditional analytics doesn't vary with time, whereas ML algorithms can enhance their performance (and accuracy) as more data is made available. For example, ML algorithms can make predictions based on the available data, compare those predictions against the actual results, and then adjust the algorithm's parameters so that they're more accurate while making future predictions.

The following figure shows a clear distinction between AI, ML, and deep learning:

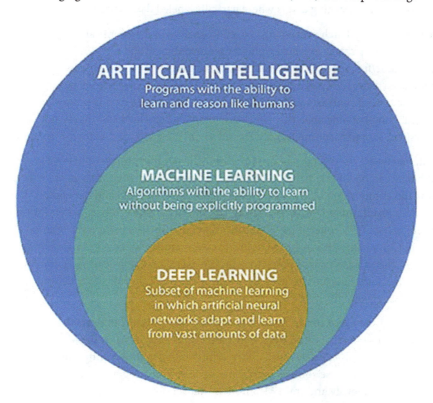

Figure 10.1 – Relationship between AI, ML, and deep learning

- **Supervised and unsupervised learning**: Supervised learning is used in AI/ML in scenarios where labeled data is available for training models. Its objective is to develop a function that can map an arbitrary input value to an output value after the algorithm is fed with the required set of input and output value pairs.

In the case of unsupervised learning, there is no need for labeled data. This contrasts with supervised learning because, in unsupervised learning, only input data is provided and algorithms must determine patterns from this input data.

Both supervised and unsupervised learning are used in IoT use cases, with the availability of the labeled data as the key selection criteria – monitoring industrial products on the assembly line for defects would require supervised learning (verification is required against a finite/ known set of defects/anomalies), whereas navigating an autonomous tractor would require unsupervised learning. Essentially, supervised learning focuses on learning concepts and/or predicting values based on historical data, whereas unsupervised learning is concerned more with determining structure and/or data clustering autonomously.

Another related term is **semi-supervised learning**, which falls somewhere between supervised and unsupervised learning and is where partially labeled data is used.

- **Batch processing**: Batch processing is used to analyze high-volume and repetitive data. This type of processing can be executed without user involvement and is typically scheduled to run at a specific time/frequency (when resources are relatively free). Batch processing works on static accumulated data and works by processing large sets of data in a parallel and distributed fashion. Batch processing contrasts with **streaming analytics**, where data is processed as it is being received. Long-running analytics where instantaneous results/decisions are not needed (sales predictions, shopper sentiment analysis, and so on) are good candidates for batch processing.

- **Cluster analysis**: The objective of cluster analysis is to group similar entities in a group (cluster) and hence can be differentiated from other entities that are dissimilar in some respect. The count and composition of groups are normally not pre-decided and depend on the received data. As a result, **unsupervised learning** (mentioned earlier) is best suited for performing cluster analysis. This concept has applications in areas such as pattern recognition and image analysis. Cluster analysis is required in IoT to categorize diverse data generated by field devices. Similarly, cluster analysis is required to associate operational tasks (such as anomaly detection, which refers to finding and flagging any sensor data that is outside of the expected range) and security issues with diverse device types.

- **Data lake**: A data lake stores data in its original format and can process structured, semi-structured, and unstructured data. With attributes such as centralized storage (of raw as well as refined/analyzed data), scalability, affordability, manageability, and security, it helps ingest data from diverse sources (social media feeds, sensor data, and data from enterprise systems) continuously and train ML models that serve the needs of data scientists, domain experts, operational/non-technical users, and others. Generally, a data lake is implemented in inexpensive storage and realized by creating data pipeline references that use raw data and generate pre-processed, semi-processed data, and processed data. Since most IoT use cases have a requirement to process historical as well as real-time data, using a data lake becomes a natural choice.

- **Data lineage**: Data lineage provides the visibility/traceability of the data as it moves from one stage to another in the data pipeline. It stores information such as the source of the data, the benefits of capturing/storing data, the different transformations that have been applied, and the intended consumers of the transformed data. In addition to providing visibility at the organizational level, often, **data traceability** is required to comply with regulatory guidelines. With strict guidelines from regulatory bodies for monitoring private/personal data, data lineage gains special importance as a robust mechanism that provides traceability of all data changes as it moves from one stage to another – especially when it comes to highlighting the stages where data is aggregated, anonymized/masked, and so on.

- **Data warehouse**: A data warehouse is a data repository that is designed to handle reporting and business intelligence needs and primarily stores/processes structured data. The key difference between a data warehouse and a data lake is that whereas a data lake stores both unprocessed (raw) as well as processed data, a data warehouse stores only structured and processed data. Another difference is that in the case of a data lake, the focus is on data storage, whereas in the case of a data warehouse, the focus is on how data is retrieved for analytics and decision-making purposes. In general, a data lake provides more flexibility as it maintains a copy of the raw data, which can be subjected to different analytics algorithms.

- **Deep learning**: Deep learning can be considered a subset of ML and works by mimicking the functioning of the human brain, where it continuously learns and refines the results with time. Deep learning algorithms are characterized by the presence of multi-layer architectures, where knowledge (learning) is progressively developed with each transition from one layer to another. For example, in the case of *image recognition*, the initial layer would only recognize pixels, the subsequent layer would focus on leveraging this understanding of pixels to develop concepts such as edges, contours, and more, and subsequent layers would augment this understanding of edges and contours to deduce higher-level features such as faces, hands, and so on.

The concept of deep learning, where learning is split across layers, is shown in the following figure:

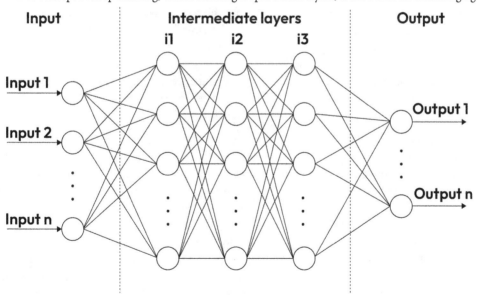

Figure 10.2 – Deep learning mimics the functionality of the human brain by processing data in layers

Deep learning algorithms analyze the available data to generate patterns that can be used for complex decision-making. Deep learning can continuously evaluate the accuracy of predictions, thereby reducing the probability of false alarms. Some of the possible applications of deep learning are speech recognition, computer vision, natural language processing, language translation, designing intrusion detection/prevention systems to guard against malicious actions, and more. For more details about deep learning, please refer to `https://aws.amazon.com/what-is/deep-learning/`.

- **Recurrent neural networks** (**RNNs**): These are neural networks that are based on algorithms that make use of sequential (or time series) information to make predictions. Most generic neural networks work under the assumption that the input and output are not co-related; however, RNNs work by considering the insights gained from prior data while analyzing the current data stream.

  RNNs are used for various applications, such as video analytics, predicting the **remaining useful life** (**RUL**) of the manufacturing asset, determining the possibility of an intrusion or security breach (detecting/preventing DDoS attacks), validating/rectifying data from field devices, and more. Similarly, RNNs can be used to generate more holistic insights as they can combine data from different sensors.

- **Regression analysis**: Regression analysis helps determine the correlation between two or more variables (how a dependent variable is influenced by a change in the value of the independent variable(s)). Multiple independent variables may be responsible for changing the value of the dependent variable and regression analysis helps determine which variables have more influence than others. Regression analysis helps predict future data points by extrapolating known correlations and can be used to optimize the overall production processes by controlling the inputs (independent variables) that are known to influence the production rate and/or quality. In IoT systems, regression analysis can be used to determine the relationship between different sensor values (how the density of vehicles impacts the value of air quality, for example).

- **Reinforcement learning** (**RL**): In an RL system, there is no defined output or goals, and the system continuously learns from the environment. An RL system understands the current context/environmental conditions and determines what actions/steps would help maximize the reward (desired objective). Often, an RL system is intelligent enough to sacrifice short-term rewards to maximize long-term rewards. RL can be contrasted with supervised and unsupervised learning, which are used for analytics purposes; RL, on the other hand, is used for determining short-term or long-term goals and objectives. In other words, RL learns to act independently instead of just learning concepts or finding patterns in the input data. RL is used in multiple ways in IoT, such as automatically routing forklifts in warehouses and autonomously moving drones in the agriculture domain. As we will see in *Chapter 11*, *Security in the IoT context*, RL can be used to monitor traffic patterns to determine and flag any malicious activity.

- **Federated learning** (**FL**): FL is an ML technique that trains an algorithm by distributing the learning logic (and associated dataset) over multiple field devices. Intermediate training results are sent to the central server and distributed analytics is performed until a result with the desired accuracy is achieved. FL has benefits over centralized ML (all the data is analyzed at a central server), such as enhanced security/privacy, reduced storage costs, improved performance, and better scalability. FL has applicability in IoT since training loads can be distributed over a large number of field devices.

- **Sensor fusion**: Sensor fusion refers to the process of implementing analytics where data from different sensors is combined to understand the situational context more accurately. For example, in the case of indoor location tracking use cases, data feeds from Wi-Fi and RFID sensors can be combined to provide more accurate location information.

  The main idea behind sensor fusion is to compensate for the weakness of one sensor type with the strength of another sensor type so that combined (*fused*) sensor readings are more accurate and useful for diverse operating conditions. Autonomous vehicles leverage data from different sensors (LiDAR, video cameras, and radar) to determine a more accurate understanding of the environmental context (as was detailed in the *Usage scenarios of sensors* section of *Chapter 9*).

- **Streaming analytics**: This refers to analyzing data as it is being ingested – that is, processing the events as they are being generated (in real time). Streaming analytics is required in use cases where accumulating data before analysis is not possible (the quality of the parts on the assembly line need to be analyzed using video feeds in real time to avoid faulty parts being accumulated). Another use case for streaming analytics is generating real-time alerts/notifications to stakeholders in case any sensor value goes beyond the expected range.

- **Tiny ML**: Tiny ML refers to the ML technologies, algorithms, and libraries that are specifically developed to run on constraint devices (for example, battery-operated devices with extremely low power requirements). This can be considered an implementation of edge analytics (refer to the *Edge analytics* section later in this chapter). In addition to conserving energy, implementing Tiny ML on a device helps overcome challenges regarding latency, bandwidth utilization, privacy/data security, and more.

- **Transfer learning**: Transfer learning is a variant of ML where a model that's trained for one problem is used as a starting point in another (but similar) problem. As an example, in image recognition of animals, the model, which has been trained to identify *dogs*, can be used as a starting point to identify *wolves* as there are a lot of similarities in terms of the appearance of both animals. This has advantages such as reduced training times and the ability to tackle situations where training data is not available for the target problem. In the IoT context, a model that's been developed to detect malicious traffic from one set of field devices (for example, a temperature sensor) can be used to detect the existence of malicious traffic from another set of field devices (for example, air quality sensors).

Important note

The **zero-shot learning** and **few-shot learning** techniques, which were introduced in *Chapter 3* (as part of the *AI/ML integration* section), are especially relevant in the context of IoT analytics.

Now that we understand the key terms that are used in analytics, let's delve deeper into how analytics can be leveraged in IoT deployments.

Implementing IoT analytics

Although this section specifies the characteristics/considerations of IoT analytics from a technical standpoint, it is worth noting that when implementing an IoT use case, domain know-how is equally important. This know-how varies vastly from one domain to another – for example, the mechanism for detecting anomalies in the agriculture domain would be quite different from the one used for detecting anomalies in the manufacturing domain. Some of the typical scenarios/use cases for which IoT analytics is used are shown in the following figure:

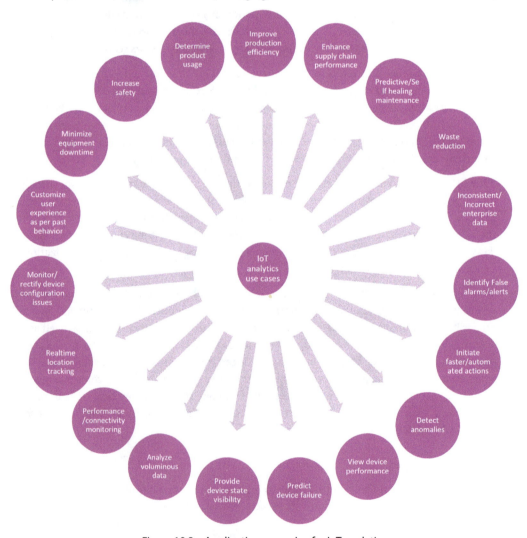

Figure 10.3 – Application scenarios for IoT analytics

IoT analytics is categorized into four different areas, depending on the insights that are generated:

- **Descriptive analytics**: In this type of analytics, stored historical data is analyzed to provide a view of historical performance, anomalies, and more. Even the real-time data stream can be analyzed, but the focus remains on the points of interest that appeared in the data stream. Descriptive analytics presents the data *as-is* (what happened and when it happened) and triggers alarms (if the data points are outside the expected range), but it does not provide reasons for data deviation/variations.

- **Diagnostic analytics**: Diagnostic analytics goes a step further than descriptive analytics and tries to understand why anomalies have been reported by field devices, observed deviation in device behavior, causes of erroneous data reported by the device, and so on. Essentially, diagnostic analytics uses data exploration techniques to establish hidden patterns and relationships between different data sources.

- **Predictive analytics**: Here, the focus is to analyze the data to provide insights into possible future outcomes. Predictive analytics helps us understand how a device will behave in the future. This is of immense value as proactive maintenance can be requested (avoiding equipment downtime) if it is known that a device will fail in the coming days. Predictive analytics applies techniques such as statistical modeling, forecasting, and ML to the output generated by descriptive and diagnostic analytics to determine possible outcomes in the future.

- **Prescriptive analytics**: This is the most advanced type of analytics, where the focus is not only on determining events of interest but also providing recommendations on how positive outcomes can be maximized and negative outcomes can be minimized. Additionally, some of the algorithms are self-learning in nature, where they can analyze the results of past recommendations to further refine future recommendations. For prescriptive as well as predictive analytics, the overall context (situational and/or temporal) becomes an important input. The system can act on the recommendations on its own or can leave the implementation of these recommendations to humans (*human-in-the-loop*). If multiple actions can result in achieving the same goal, prescriptive analytics can evaluate each of these mechanisms and advise on the optimum action path.

These four types of analytics have been exemplified using *smart agriculture* in the following figure:

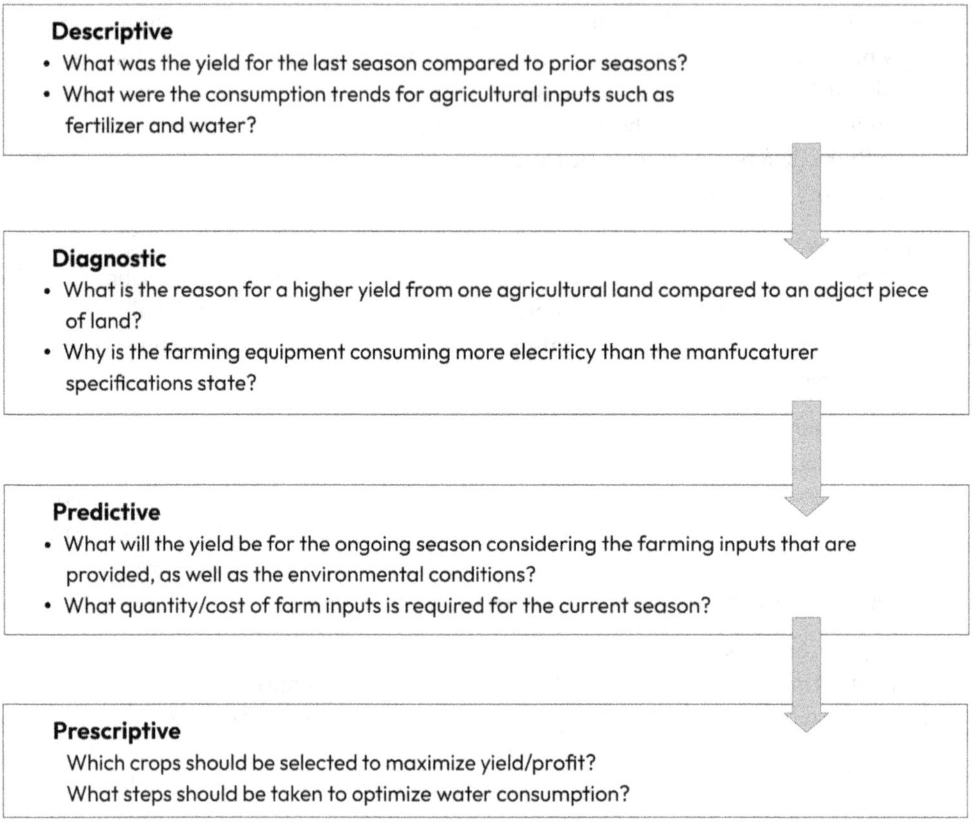

Descriptive
- What was the yield for the last season compared to prior seasons?
- What were the consumption trends for agricultural inputs such as fertilizer and water?

Diagnostic
- What is the reason for a higher yield from one agricultural land compared to an adjact piece of land?
- Why is the farming equipment consuming more elecriticy than the manfucaturer specifications state?

Predictive
- What will the yield be for the ongoing season considering the farming inputs that are provided, as well as the environmental conditions?
- What quantity/cost of farm inputs is required for the current season?

Prescriptive
Which crops should be selected to maximize yield/profit?
What steps should be taken to optimize water consumption?

Figure 10.4 – Types of analytics in the smart agriculture context

In addition to **batch analytics**, where data is analyzed *post factum* for long-term trends/insights, in IoT use cases, you need to analyze data in real time (streaming analytics) so that timely corrective actions can be taken. For example, detecting and resolving a quality issue on the assembly line in real time is much more beneficial than being aware of the issue at the end of the shift. Also, as data is being analyzed in real time, resending erroneous/faulty data after corrections from field devices to the central server is not practical.

For a holistic analysis, the data from sensors needs to be combined with additional data sources. Examples of this include **enterprise resource planning (ERP)** and a **human resource management system (HRMS)**. For details, refer to the *Enterprise system integration pattern* section in *Chapter 3*. Data may need to be merged from multiple sensors with specified enterprise systems to arrive at more accurate decisions/comparisons. Consider the case of a retail store where sensors are placed to determine the areas that are frequented by shoppers. Richer insights can be obtained if this data is correlated with sales data to determine areas that are resulting in actual sales and if the recent changes in the store's layout resulted in an increase in sales. Also, combining feeds from multiple sensors, such as video cameras (for identifying the shoppers) as well as RFID readers (for identifying items in people's shopping carts) will give more accurate insights into shoppers' purchasing behavior.

IoT sensors are often installed under extreme conditions (high temperature/humidity, subject to dust/vibrations, and so on), which can result in random data losses or issues regarding incorrect/duplicate data. Since these devices are constrained, it is difficult to perform operations such as data sanitization and data deduplication on the source device and shift the implementation of these data cleanup activities to the central server.

Analytics can be performed on both operational and diagnostics data sent by the sensors. Operational data is related to the actual data that's captured by the sensor (ambient conditions), whereas diagnostics data is used to determine the current (or predict the future) health of the device/sensor. Due to the constraints mentioned earlier, one type of data may be prioritized over the other – being able to transmit operational data is suspended until the device's health is back to normal (for example, worn-out batteries have been replaced).

Analytics can also be used for promptly detecting and mitigating security incidents. The analytics algorithm learns the normal/usual data traffic patterns and then monitors the traffic patterns for any anomaly indicating a potential security incident. Analytics can be used to inspect individual packets to determine whether the contents of the packet have been modified during transit or to determine and remediate the possibility of malware or device misconfiguration(s).

In the next section, we'll look at the key stages involved in IoT analytics.

Stages of implementing IoT analytics

IoT analytics workloads generally go through a series of steps, which include aggregation, pre-processing, transformation, and processing, with the final stage being to present the analytics results to the user, as shown in the following figure:

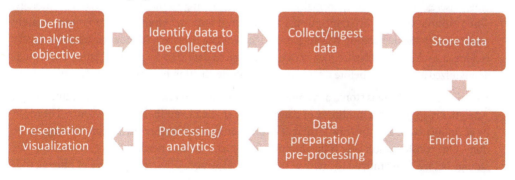

Figure 10.5 – Different stages of IoT analytics

Let's look at these stages in more detail:

1. **Defining the analytics objective**: This stage includes understanding the business imperatives and business processes and defining the information needs of the users, and the goals/objectives that the analytics program seeks to accomplish. Stating the business problem clearly helps align everyone on the expected result/benefits of the analytics initiative. Some examples of the goals/ objectives that can be possible outputs at this stage are listed here:

    *   Is it possible to increase production output without compromising on quality?

    *   In the agricultural domain, what agricultural inputs should be modified to increase yield?

    *   In terms of our retail store example, how should the layout of the retail store be modified to improve *shopper footfall* or *sales per square foot*?

2. **Identifying the data to be collected**: This step involves identifying the type of data to be collected. This includes primary data that may be collated from various sensors/devices and secondary data that can be derived from primary data by applying a set of mathematical and statistical operations on the primary data. In addition, secondary data can be obtained from enterprise systems or open datasets available in the public domain. This step might reveal the need to install additional sensors to capture the required data. If real sensors can't be deployed for any reason, how the data stream would be simulated/fabricated (virtual sensors) is also finalized in this step.

    The type, frequency, and amount of data that's collected varies depending on factors such as the desired accuracy of the analytics results, the time criticality of the data being collated (any reading beyond the defined delay threshold is not relevant and is discarded), and the energy budget of devices/sensors (collating and transmitting data consumes energy – the higher the frequency of data capture/transmission, the higher the energy consumption is).

3. **Collecting/ingesting the data**: This step also involves obtaining data from identified sources, as well as developing connectors to fetch data from internal/external data sources. Data from enterprise/external systems is normally collected using **change data capture** (CDC) technology, where source systems are continuously monitored and any changed data is immediately transferred to the destination (via an event-triggered data transfer). In other cases, data transfer is done at a scheduled time/frequency (a time-triggered data transfer). Often, data is encrypted/ anonymized at this stage before it's transmitted to its destination to avoid privacy/security issues.

4. **Storing the data**: Data is stored at a common location for immediate processing or future usage. Analytics is applied to a multitude of data sources, so data from these diverse sources needs to be accumulated in a native/raw format at a common location before further processing can be done. One consideration here is the duration for which raw data needs to be preserved in active storage before being moved to archival (less costly) storage. Often, the process of moving old/ stale data is carried out in an automated fashion based on the configured rules.

5. **Enriching the data**: Here, data from different sources is combined to obtain a more holistic dataset. Combining dynamic (telemetry data from a sensor) data with static data (device metadata such as device location or serial number) is a typical enrichment in the IoT context. Sensor data received from field devices would be enriched by data available in enterprise systems (ERPs, CRMs, and so on) or external/third-party data sources (weather data).

    Data generated by field devices is typically spatiotemporal and needs to be combined with additional context/metadata so that it can support analytics use cases. Consider a retail store scenario – RFID readers/sensors would point to the current location of a grocery item, but to pinpoint the items that are past their usage/expiry date, data from additional systems would be needed.

6. **Preparing/pre-processing the data**: This step involves removing outdated, noisy, and duplicate/redundant data, checking for and reformatting incompatible data types, and filling in missing data. Missing values are replaced with probable values by using adjacent readings and/or by understanding the overall context. This step also includes dividing the data into two sets – one for training the ML/deep learning model and another for verifying the accuracy and robustness of the developed model.

    This step can include **data anonymization** if data privacy is a requirement and may also involve transforming standard data formats. While performing these transformations, a copy of the raw data is also stored for future needs.

    The data preparation/pre-processing step is essential for improving the quality of ingested data as insights provided by any analytics algorithm would be determined largely by the quality of the data that's fed into the analytics algorithm.

> **Important note**
> Data preparation is a crucial task and may consume up to 70-80% of the time required to execute the complete data pipeline.

7. **Performing data processing/analytics**: Data is subjected to selected analytics algorithms generating the required insights, decisions, or predictions. The type of analytics algorithm that's employed will depend on the type of analytics (descriptive, diagnostic, predictive, or prescriptive) that is required. This step also involves using data mining tools for ML model training and deployment, clustering/classifying pre-processed data, and generating results in the form of trends, data summaries, anomalies, predictions, recommendations, and more. In the case of streaming analytics, automated actions are also triggered at this step, as per configured rules.

8. **Visualizating/presenting the data**: This step involves presenting the analysis results in the form of dashboards or reports so that information can be consumed by both technical and business users. This step also allows users to compare datasets, observe relationships, and arrive at conclusions. The output of this step will determine whether the goals/objectives that were set for the *data analytics* initiative have been met or not. This is the stage where the real benefits/value of implementing an IoT use case are realized by stakeholders such as the project/product team, project sponsors, and end users so that they can derive actionable insights.

Advancements in AI and natural language processing have enabled users to state their information or analytics needs in everyday language and get the results in conversational format (this will be covered in detail in the *Large language models* and *Generative AI* sections in *Chapter 12*).

Visualizations that use **augmented reality** (**AR**) technology, where analytics results are overlaid on a field view in real time, are another useful advancement in recent times:

Figure 10.6 – Displaying analytics results using augmented reality

This step is important as it decides what results need to be displayed and hidden (to avoid information overload). Based on the information presented in this step, decisions can be made and actions can be initiated. One possible decision can be to execute further/deeper analysis by executing more analytics cycles – that is, restarting the analytics process from *Step 1* onward.

Now that we know about the various stages involved in IoT analytics, let's look at how these processes can be enhanced.

Integrating ML capabilities into IoT analytics

ML is a subset of analytics that allows systems to automatically learn from past data without being explicitly programmed. Adding ML capabilities allows IoT systems to automatically detect patterns and anomalies. One challenge in developing ML models for IoT use cases is the non-availability of training datasets. Most of the data that's generated is specific to the particular use case and is highly dependent on the sensor and its operating conditions. Since these conditions are extremely difficult to simulate, getting a *standard* dataset for IoT use cases is difficult. Also, the data would be labeled (in the case of supervised learning) by a person who has deep domain knowledge. Such domain experts are not easily available or are expensive to hire.

As a result, transfer learning is typically used for developing models. Here, a pre-trained model is used as a base and then fine-tuned using application-specific data. Tiny ML can also be used, which refers to techniques that are used to run ML algorithms on constraint devices, especially low-power and battery-operated devices.

The purpose of analytics is to highlight outliers/anomalies and find correlations between data generated by multiple sensors and data sources. In addition to application use cases such as predictive maintenance and automatic anomaly detection and correction, IoT analytics (and specifically ML) can also play a key role in designing self-calibrating and self-healing field devices.

Domain experts need to be involved when designing ML models; however, domain expertise is required continuously for model validation and refinement. This results in a continuous interplay between humans and machines, as shown in the following figure:

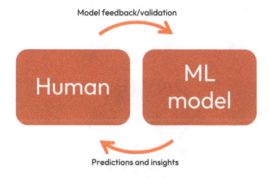

Figure 10.7 – Interplay of human feedback and ML model refinement

Another reason why domain expertise needs to be *codified* in ML algorithms is related to societal changes – for example, a considerable number of employees with domain expertise are in the process of being retired with very few replacements (this situation is particularly alarming in the manufacturing sector). In some scenarios, the human mind might not be able to make sound rational decisions or judgments due to preconceived notions or inherent cognitive biases (confirmation bias, attribution bias, availability bias, and so on). Here, AI/ML algorithms can be used to neutralize such biases – provided the training process/training data has been closely monitored/scrutinized to ensure it can't be impacted by possible biases.

To effectively implement analytics, organizational process flows (workflows) need to be examined, with a special focus on decision-making steps. Some of these steps can be automated using IoT analytics. Analytics can identify and recommend how you can automate the steps that are causing performance bottlenecks and hence expose system-level inefficiencies.

This section covered the different types of analytics that are relevant in the IoT context. However, the results that are generated by analytics are only accurate if the input data is correct and high-quality. In the next section, we will talk about the factors that impact the quality of data and the mechanisms that can be used to improve it.

Understanding the importance of data quality

The different types of data quality issues that are generally found in IoT use cases can be seen in the following figure:

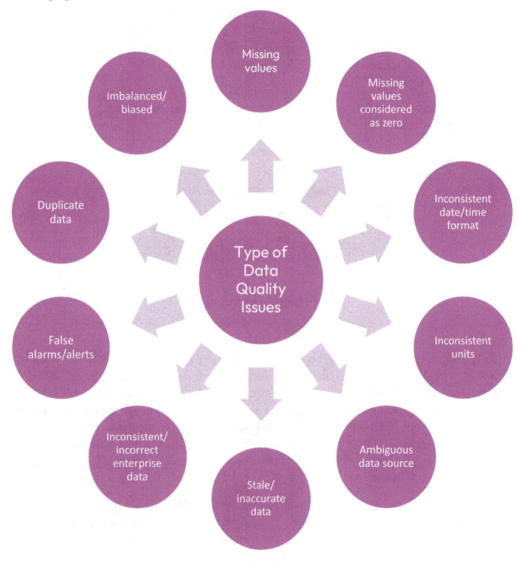

Figure 10.8 – Typical data quality issues

The source of these data quality issues can be traced to different layers of the IoT reference stack, as shown in the following figure:

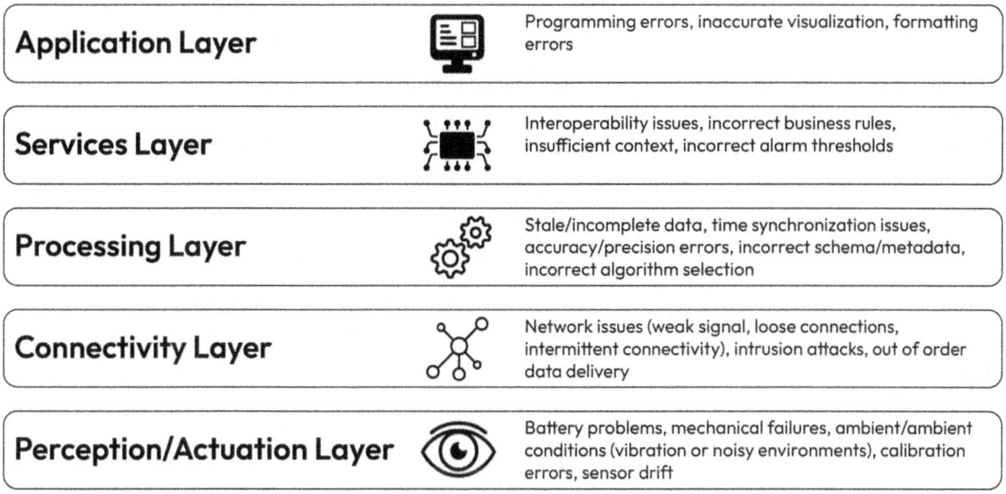

Application Layer		Programming errors, inaccurate visualization, formatting errors
Services Layer		Interoperability issues, incorrect business rules, insufficient context, incorrect alarm thresholds
Processing Layer		Stale/incomplete data, time synchronization issues, accuracy/precision errors, incorrect schema/metadata, incorrect algorithm selection
Connectivity Layer		Network issues (weak signal, loose connections, intermittent connectivity), intrusion attacks, out of order data delivery
Perception/Actuation Layer		Battery problems, mechanical failures, ambient/ambient conditions (vibration or noisy environments), calibration errors, sensor drift

Figure 10.9 – Data quality issues at different layers of the IoT stack

The scale of IoT deployments (for example, a large number of field devices generating humongous data) tends to amplify even minor quality issues. The tolerance for data quality issues varies across organizations and use cases, and accordingly, the rigor of data quality mechanisms will also vary.

Applicable data quality initiatives will depend on factors such as the nature of the data collected and the purpose for which the data is being collected.

Closely related to the concept of data quality is the concept of *data lineage*, where the traceability of data is maintained as it moves from one stage to another in the data pipeline – that is, it captures the flow of data from its source to its destination, as well as any changes that are introduced as it moves from one stage to the next. Data lineage ensures data quality as it provides visibility into the stage at which change (or a quality issue) was introduced and the nature of that change. This information is crucial in developing data quality workflows where automation tools can detect and fix quality issues without manual intervention.

This section covered the different types of data quality issues that are encountered in IoT cases and the factors that result in data quality issues. The next section will introduce edge analytics, where data processing is done closer to the point of data generation.

Relevance of edge analytics

IoT devices are not permanently connected to a central server, so some amount of processing/analytics needs to be done locally so that these devices can function independently if they're not connected to a central server. This is one scenario where *edge* analytics is required. Essentially, edge analytics refers to processing IoT data near the point at which it is generated. In other words, edge analytics refers to the scenario where analytics data is sent to the point of data generation rather than being sent to the point where analytics and algorithms are hosted or deployed. This definition points to the fact that edge analytics can be implemented on a variety of physical infrastructures (device gateways, on-premises servers, or data centers physically located close to field devices).

Distributing data processing workloads between edge and central server depends on use case requirements – most IoT use cases rely on a hybrid approach. Usually, latency-sensitive data is processed at the edge and data that requires heavy compute/storage is processed at the central server. A connected car is a good example of this hybrid approach.

Edge analytics is also required in scenarios where data security and/or privacy is a concern. Privacy/security concerns limit the possibility of sending data to the central server over public communication channels (the internet). Pre-processing data (filtering, aggregation, downsampling, and so on) at the edge helps conserve the communication bandwidth, improve data quality (by reducing noisy/redundant data) at the source, and optimize the power consumption of field devices (data transmission is a major source of power consumption).

Field devices may generate huge amounts of data, but very few data points are of relevance – that is, data points that indicate anomalous/erroneous conditions. Therefore, it is important to filter out the redundant data using edge analytics rather than first transmitting data and discarding it at the central server level. One use case that shows the relevance of edge analytics is *video analytics* as there is no need to send the (bandwidth-hogging) video streams to the central server when these can be analyzed at the edge, where the result (presence or absence of the object of interest) is sent to the central server.

As mentioned earlier, the proportion of processing that's done at the edge or at the central server will vary based on use case requirements, as well as the technical capabilities of physical infrastructure. Distributing processing between the edge (local) and central server (global) can be visualized as a continuous scale, with complete processing at the edge on one end and complete processing at the central server on another end, as shown in the following figure:

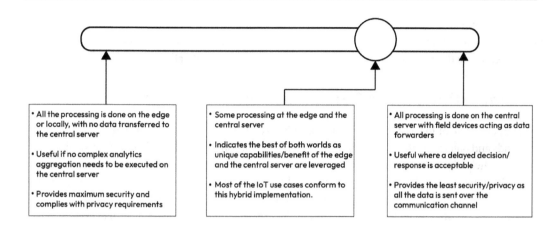

- All the processing is done on the edge or locally, with no data transferred to the central server
- Useful if no complex analytics aggregation needs to be executed on the central server
- Provides maximum security and complies with privacy requirements

- Some processing at the edge and the central server
- Indicates the best of both worlds as unique capabilities/benefit of the edge and the central server are leveraged
- Most of the IoT use cases conform to this hybrid implementation.

- All processing is done on the central server with field devices acting as data forwarders
- Useful where a delayed decision/response is acceptable
- Provides the least security/privacy as all the data is sent over the communication channel

Figure 10.10 – Data quality enhancements at different stages of the data pipeline

Traditionally, analytics was done centrally at the central server as limited intelligence (and compute) was available for field devices. Limited intelligence helped keep the architecture of the field devices simple, thus improving their reliability/longevity. **Reliability** is the prime requirement as devices are normally deployed at remote locations where any repair/replacement is difficult. Adding more intelligence to the field devices resulted in a corresponding increase in complexity and reduced reliability (more code logic at the edge, thus increasing the probability of more defects). It was much easier and more efficient to fix defects at the central server rather than releasing patches to devices spread out in the field. As a result, most of the analytics was performed on the central server. However, with the increase in hardware/software capabilities (Moore's law), the *edge* can execute relatively complex algorithms. This explains the transition to move processing from the central server to the edge. Edge processing also increases the system's overall scalability as the central server's load is spread among a large number of edge devices.

The nature of the analytics that's performed at the edge should be carefully considered as performing more analysis results in less raw data being available for future analysis. Some applications circumvent this limitation by storing a copy of the raw data at the edge and then transferring it to the central server after a specific event occurs. An example of this is when a field device comes physically close to the central server and the mobile device gateway enters the connectivity range of the central server. Data stored on the central server should be carefully curated to avoid the possibility of a **data swamp**, where data is dumped without any clear purpose.

As edge analytics is performed in constrained environments (field devices have limitations regarding power/compute/storage and communication/connectivity challenges), generic algorithms need to be adapted to work in such environments. In summary, edge analytics should be composed of lightweight, purpose-built, modular, and resilient components.

This section defined edge analytics, how it differs from analytics performed at the central server, and the scenarios where it can be leveraged effectively. Analysis results need to be displayed to the user so that they can take the required actions. We will look at this in more detail in the next section.

Considerations for IoT visualization

The main objective of IoT visualization is to make it easier for information consumers to understand data trends and obtain insights by highlighting patterns, relationships, trends, and outliers. Visualization should also help disseminate insights to non-technical users. The rationale regarding the decisions or actions taken by AI/ML algorithms is generally hidden from the user. Effective visualization can also help fill this gap by providing transparency into how AI/ML algorithms have arrived at a decision. Similarly, data lineage can be better understood if it is represented in the form of visualizations.

IoT data and insights are consumed on a diverse set of devices. In addition to traditional devices such as desktops and mobile devices, information is consumed on **human-machine interface** (**HMI**) devices (such as industrial control panels with buttons and indicator lights). This requires entirely different layout considerations than those that are relevant to designing a desktop application:

Figure 10.11 – IoT data being monitored via an HMI device

Here are some factors that determine the type of visualization that's used:

- **Charts/dashboard elements**: Some of the representative options include the following:

  - Scatterplots (show relationships)

  - Interactive maps (provide geospatial information)

  - Bar charts (show a comparison between two or more data types)

  - Pie charts (show distributions)

  - Line graphs (display changes over a date/time range, time series data)

- **Need for displaying real-time data updates**: The number of data points that correspond to the parameter(s) shown in real time would be large. This necessitates the need to highlight anomalies (or points of interest) with different colors, markers, and so on. Similarly, alerts/notifications where action is needed from the user should be made visible via popups.

- **The information needs of the users**: Highly technical data with a minimal business context might not be appreciated by business users. One way to design a visualization interface is to start with objectives-based design – what decisions/conclusions does the user need to make based on the data being displayed?

- **The type of customizations that are allowed**: Some dashboards allow users to control the type and amount of data that is transmitted by field devices. Visualizations also provide a *self-service* capability, where users can configure the displayed content (level of detail) and layout (data/insights ordered by importance) as per their preferences.

- **Data filtering**: Options for data sorting/data filtering to help users sift through large amounts of data.

- **Level of granularity that is allowed**: Most dashboards provide information at a high level and an option to drill deeper into trends of interest.

- **Priority of information messages**: Higher priority messages/notifications shouldn't be hidden by lower priority messages. To do this, you can visually segregate normal events from anomalies by using different colors.

This section covered the relevance of visualization in IoT use cases and factors that need to be considered while designing visualization for IoT use cases. Now, let's summarize what we've learned so far.

Summary

This chapter covered the importance of analytics in the context of IoT. This process converts raw data that's received from diverse sources (static data sources such as enterprise systems, as well as dynamic data sources – for example, data received in real time from sensors) into meaningful insights, which is a prerequisite for effective/efficient decision making. We covered specific considerations that you need to be aware of while tailoring *generic* analytics for IoT applications. Then, we covered the steps that are normally followed in any analytics data pipeline. We also looked at the benefits and nuances of edge analytics, where data is processed close to the data source.

After that, we introduced edge analytics; we will cover this in more detail in *Chapter 12, Exploring Synergies with Emerging Technologies*. We looked at the importance of data quality and how it can be ensured at the different layers of the IoT stack, as well as the importance of IoT visualizations.

This chapter should have helped you understand general analytics concepts and how they relate to IoT use cases. You were also introduced to some techniques/concepts that are especially relevant to IoT, including sensor fusion and HMI. Equipped with this knowledge, you should be able to develop IoT solutions by selecting the right analytics techniques for the problem at hand.

In the next chapter, we will cover an important topic known as **IoT security**. The decentralized nature of IoT deployments and the constrained nature of field devices results in a unique set of challenges for providing end-to-end security to IoT workloads. We will look at how these challenges/constraints are handled, as well as possible security vulnerabilities (and their mitigations).

11
Security in the IoT Context

Previous chapters provided examples of how IoT can be used to provide solutions to achieve various business targets within diverse problem domains. As a solution scales and graduates from the **proof of concept** stage to the **production deployment** stage, one crucial barrier that needs to be crossed is making these solutions secure from the acts of malicious actors. **Security** generally has the highest priority among **non-functional requirements** (**NFRs**) in any solution; however, in the case of IoT solutions, this requirement becomes non-negotiable due to the potential risk of material and/or human loss. It is no surprise, then, that security is cited as one of the topmost factors that can limit the adoption of IoT solutions at a large scale.

Non-existent and patchy regulatory security norms are among the reasons why consumers often make do with solutions released with security vulnerabilities that are exploited by malicious actors on the first day of solution deployment (**zero-day vulnerabilities**). The attack surface of IoT solutions is much broader (a large number of field devices as well as traditional IT systems such as a central server) and deeper (vulnerabilities exist at all layers of the IoT stack), so traditional approaches to IT security aren't sufficient. However, it is important to mention here that several countries are actively working to formulate cybersecurity guidelines (e.g., the Cyber Resilience Act – applicable to European Union countries) to contain potential vulnerabilities/threats. Going forward, it is likely that the implementation of these guidelines will be made mandatory for products to be launched in the respective geographies.

To implement a robust security posture, IoT architects build upon the knowledge gained in securing traditional IT systems and add additional safeguards/guardrails that are relevant to IoT/OT solutions:

Figure 11.1 – IoT security combines elements from IT as well as OT security

This chapter presents the security vulnerabilities that are observed at different layers of the IoT stack as well as their mitigations. Although the focus of the chapter is listing the vulnerabilities (and their mitigations) inherent in field devices (perception/actuation layer of the IoT stack), this is not to say that vulnerabilities aren't present at other layers of the IoT stack. The reason for focusing on securing field devices in this chapter is that the remaining layers are well covered in *generic* IT security literature.

Before delving deeper into the vulnerabilities present in IoT solutions and their mitigations, let us first understand the key terms and definitions that are frequently used in IoT security literature.

Key terms/definitions

In this section, we'll discuss some of the key concepts in the field of IoT security:

- **Air-gapped systems**: Having an air-gapped system means not connecting two or more networks unless there is a specific need for connection and it is implemented primarily to reduce the probability of security attacks – a threat actor can't use the vulnerability in one network to target other connected networks.

- **Asymmetric encryption**: Asymmetric encryption uses a public key for encryption and a private key for decryption, unlike symmetric encryption, where the same key is used for both encryption and decryption. Asymmetric encryption requires more computational power than symmetric encryption but provides more robust security.

- **Attack vector**: Attack vector refers to the method and/or mechanism used by threat actors to compromise the security of the system.

- **Authentication**: Authentication refers to the ability of the receiver to verify whether the data sent by the sender originated from a trusted/known source. Without the ability to authenticate the data, threat actors can insert spoofed data (sensor reading, control commands, configuration data, firmware packets, and so on) into the communication channel.

- **Blast radius**: Blast radius refers to the extent or impact of the adverse security event. For example, in the IoT context, we can say that the blast radius of the field device getting compromised is much smaller than the scenario where the central server is compromised.

- **Botnet**: A botnet attack is orchestrated by infecting malware in a group of IoT devices and then instructing them to send simultaneous requests to the target server. IoT devices serve as a good launchpad for botnet attacks due to their weak security, large scale, and similar traffic patterns from most of the field devices.

- **Command injection**: Command injection implies a vulnerability where a remote authenticated user can execute desired commands on the field devices or the central server. Oftentimes, the malicious commands are disguised as valid system commands. Providing strong end-to-end encryption and integrity checks can help mitigate this vulnerability.

- **Certificate Authority (CA)**: A CA is a trusted entity that issues and validates certificates and is trusted by both the certificate owner and the party using the certificate.

- **Certificate rotation**: Certificate rotation refers to the process of replacing expired, invalid, or compromised certificates with newer and valid certificates. In normal circumstances, certificates should be rotated at regular intervals and well before the certificate's expiry date.

- **Credential bootstrapping**: This refers to the initial exchange of credentials to set up communication between devices or between the device and the central server.

- **Confidentiality, Integrity, and Availability (CIA)**: CIA (often referred to as the CIA triad) refers to the fundamental tenets of confidentiality, integrity, and availability upon which all the security mitigation approaches are based:

  - **Confidentiality** refers to the protection of sensitive information (data providing a competitive advantage, device certificates, passwords, and security keys) from unauthorized access both at the time of storage as well as while data is in transit. Also, the data should be accessible only to authorized persons.

  - **Integrity** means protection against the illegitimate modification of data, indicating the capability of the central server to receive accurate/correct data from the sensor, send desired/required commands to actuators, or securely transfer firmware updates to field devices without them being tampered with. Integrity is ensured by using checksums, hashing, and digest algorithms. These integrity checks are performed before accepting data over a communication channel as well as during the device's boot-up sequence. Depending upon the level of sophistication required, some field devices might check the validity of the firmware/software during actual execution as well.

  - **Availability** refers to the requirement that data should always be available to authorized users and if it isn't available, interested parties should be notified beforehand. The availability of IoT services or devices can be disrupted by flooding them with high traffic – much more than the intended load. The unavailability of IoT services typically results in financial loss to the service providers and functional loss to the users.

  The importance of each tenet of CIA differs from one use case to another and an understanding of the trade-off involved in prioritizing one over the other is the basis of designing mature security solutions. In general, IoT use cases tend to prioritize integrity and availability over confidentiality. This contrasts with traditional IT systems where confidentiality is prioritized over integrity and availability as IoT systems tend to operate in and impact the real/physical world – loss of confidential information in a financial application (credit card details) can cause considerable financial loss, whereas a hacker can cause more collateral damage by manipulating commands in a nuclear facility (e.g., a command to initiate an uncontrolled chain reaction) – although knowledge of an actual command, per se, won't result in any major impact.

- **Cryptojacking**: Cryptojacking involves leveraging hardware resources to mine online currencies. This attack can target both the central server as well as field devices. Although individual field devices are not computationally powerful, threat actors typically distribute the load across a large number of devices for effective mining. In addition to the direct impact on performance, this might increase infrastructure costs (e.g., cost escalation due to higher bandwidth consumption).

- **Distributed Denial of Service (DDoS):** This attack involves forcing the IoT application to either stop functioning or function with degraded performance levels and is achieved by flooding the central server with a huge amount of traffic. In this kind of attack, the objective is not to corrupt/misuse the data but to render the IoT system non-operational by denying response to legitimate requests, e.g., denying system availability.

  Another variant of DDoS involves flooding the battery-operated field devices with data/command requests, thereby draining the battery. Monitoring the network traffic for abnormalities, using firewalls to block suspicious traffic, allowing devices to send/receive specific commands/instructions, and quarantining infected devices are some of the mechanisms for containing DDoS attacks.

- **Transport Layer Security (TLS) and Datagram Transport Layer Security (DTLS):** TLS is the transport security protocol for TCP traffic and DTLS is the protocol for UDP traffic.

- **Eavesdropping:** An eavesdropping attack occurs when a threat actor intercepts and/or modifies data transmitted between two devices or between field devices and the central server. Eavesdropping can be mitigated by making the data transmission secure/encrypted and by verifying the integrity of the received data.

- **General Data Protection Regulation (GDPR)** mandates that the user is aware of what personal information is collected, the purpose of data collation, and where (and for how long) data would be stored and processed. The regulation also allows users to request to delete their personal data at any time. In addition to obvious personal data (such as age, name, and gender), GDPR also covers operational data gathered by IoT devices, such as audio/video recordings, biometrics, and location.

- **Ransomware:** Ransomware refers to the malicious attempt to lock (encrypt) the user's data unless the demand (generally fund transfer) is met. In IoT applications, in addition to blocking the data, tactics such as disabling/reducing the device functionality and stealing personal data from devices are also used to force the user to agree to the unjust demands of the threat actor.

- **Least privilege access:** Least privilege access refers to the practice of granting the least (or just enough) access for the requested asset/resource. If followed consistently throughout the system, this can help to reduce the blast radius and limit the damage that a threat actor can inflict.

- **Man in the middle (MITM):** In this type of attack, the threat actor stealthily intercepts the transmitted data or commands and then replays them (often replacing the valid messages with malicious messages).

- **Mutual authentication:** Mutual authentication refers to the process where entities that want to communicate are required to verify the origin and integrity of each other before sharing data. In IoT systems, the device needs to authenticate the central server and vice versa.

- **Nonrepudiation**: Nonrepudiation indicates that the entity generating the data shouldn't be able to deny that they were the source of data – that is, they are not able to repudiate the data origin (this is similar to the case where the presence of a valid signature can't be later repudiated/denied by the signatory). Nonrepudiation can be achieved by ensuring that every logged event can be individually traced to the source and independently validated.

- **Public Key Infrastructure (PKI)**: PKI was regarded as a *gold standard* for implementing IT security in the past and provides protection against possible CIA vulnerabilities with features such as mutual authentication, firmware validation, data encryption, and integrity assurance. PKI eliminates the need for passwords by using digital certificates (based on asymmetric encryption).

- **Role-Based Access Control (RBAC)**: An IoT application should be designed to only allow those features/functionalities that are relevant/required for a particular role. This aligns with the *principle of least privilege*. A simple example can be that a person should be able to view data related to devices for which they are responsible and not others. However, the solution provider should be able to view data from all the devices (in anonymized form) for product refinement and/or troubleshooting purposes.

- **Root of trust**: The root of trust is a component on a device and is made up of highly reliable hardware, firmware, and software components that are involved in executing critical security functions. Typically, it is implemented using immutable hardware. The root of trust on a device determines the level of confidence in the authenticity of the device credentials.

- **Trusted Execution Environment (TEE)**: The TEE allows for code execution in a trusted environment and trust is achieved by providing robust isolation and restricted access to the execution environment.

- **Trusted Platform Module (TPM)**: TPM is a hardware chip that executes cryptographic algorithms and stores cryptographic keys.

- **Threat modeling/analysis**: Threat modeling is the process of determining the possible vulnerabilities in an IoT system and how they can be exploited. Threat modeling generally starts at the solution design/architecture phase and continues throughout the life cycle of the product. Threat modeling helps to identify and prioritize security-related risks and is typically done in three steps:

  - Identification and documentation of assets, components and their interfaces, actors, and privilege levels for all the workloads under consideration

  - Developing a prioritized list of possible threats (generally with the associated impact and probability of occurrence) for each of the identified asset, component, and actor combinations

  - These threats are identified using the **STRIDE model**, where **STRIDE** stands for **Spoofing, Tampering, Repudiation, Information Disclosure, Denial of Service, and Elevation of Privilege**

  - Listing possible mitigation and control strategies for each of the identified threats

- **Secure boot/measured boot**: Secure boot refers to tools and techniques used to integrate security into a field device's boot sequence by validating every stage of the boot sequence (transition to the next stage is possible only if the previous stage has been successfully validated). This ensures that the device only boots if the boot image is trusted by OEM. If there is an error at any stage of the boot sequence, a *graceful* transition is made to the previous secure state with all the residual data/code removed. This prevents threat actors from installing malware or making any other firmware changes to the field devices and is accomplished by using techniques such as image checksums and signature verification to ensure that the boot image is received from a trusted source. This requires the presence of TPM on the field device.

  Measured boot refers to the process of storing unique hash values during the boot process in TPM. These values can later be used to validate the execution sequence of the boot process and to flag any possibilities of tampering with the boot process.

- **Side-channel attacks**: Side-channel attacks allow for the extraction of sensitive information (e.g., encryption keys and configuration data) by indirect means, such as by monitoring network communication, the device's power consumption levels, changes in the device's operating temperature, or changes in electromagnetic radiations emitted by the device. Often, side-channel attacks are triggered by operating the system outside its normal state/behavior (where it isn't sufficiently tested by the manufacturer) through tactics such as placing the device in an environment with high electromagnetic radiations or operating it in a non-recommended environment (extreme temperature/humidity conditions). Some mechanisms that can be used against side-channel attacks include thorough testing of the system (especially abnormal operating conditions), obfuscating the device's electric signals by introducing dummy operations, housing devices in a protective electromagnetic shield/enclosure, and ensuring that the device circuitry fails gracefully if operated outside its normal operating conditions.

- **Security Information and Event Management** (**SIEM**): SIEM helps with the real-time monitoring and logging of security events. It helps to proactively identify and mitigate possible system vulnerabilities by finding and correlating patterns. SIEM uses **artificial intelligence/ machine learning** (**AI/ML**) techniques to flag anomalies and generate notifications/alarms to the relevant stakeholders. Data/insights generated by SIEM can also be used for auditing or regulatory compliance purposes.

- **Zero-day vulnerability**: Zero-day vulnerabilities refer to the potential vulnerabilities or flaws that are present on the day of solution release/deployment (zero day) and are not known to the solution providers but can be exploited by threat actors.

- **Zero-trust network**: As the name suggests, this refers to fully securing each component in the IoT system without making any assumptions about the security capabilities of the individual components. Traditional security approaches rely on network segmentation as the prime defense against security exploits where devices within a network are considered trustworthy. However, *zero trust* requires that entities always prove their trustworthiness via secure authentication.

- **Intrusion Detection System (IDS)/Intrusion Prevention System (IPS)**: An IDS monitors the observed events against a set of known vulnerabilities/intrusions and/or by determining the normal/baseline behavior and then flagging anomalous/outlier events. The key difference between an IDS and IPS is that an IDS is limited to sending alerts/notifications if anomalous behavior is detected, whereas an IPS is capable of initiating required remedial actions as well. The agent for collating event data is installed on field devices, but the analysis/processing required for an IDS/IPS is generally done at the central server due to complex computation requirements.

- **Vulnerability assessment**: A vulnerability assessment should be done at the start of the deployment and then at regular intervals to ensure that the system is able to identify and mitigate any new security vulnerabilities.

Now that we have a reasonable understanding of security terms, let us understand how implementing IoT security is quite different from traditional cybersecurity (or IT security).

Comparing IoT security and IT security

IoT security aims to provide different tools, techniques/mechanisms, and strategies for identifying, monitoring, and controlling IoT vulnerabilities. It is difficult to provide security to IoT solutions compared with IT solutions due to the following reasons:

- The **number of devices** that are connected to a network is much higher in IoT networks than in IT networks. Also, since a large number of devices needs to be deployed, devices that are chosen for deployment are inexpensive to keep the overall cost low. This results in devices that have low compute, storage, and power capabilities that can't execute complex (albeit robust) security algorithms. Field devices are often deployed in remote locations with intermittent connectivity, which hinders the ability of these devices to receive timely security updates/patches.

- Another related aspect worth considering is the fact that traditional IT systems are well protected by **physical/perimeter security**. However, the same is not true for IoT field devices. Threat actors can gain physical access to these devices and understand the internal functioning and/or communication protocols used by these devices. The knowledge gained from one device can then be used to exploit vulnerabilities of other similar devices.

- IoT solutions are designed to **operate independently** and without human involvement. As a result, they rely on security mechanisms that differ from normal IT systems (certificate-based authentication used in IoT solutions compared with username/password mechanisms in IT systems).

- The **impact of security attacks** in an IoT space is more severe than IT security attacks as human life and/or critical infrastructure are the target. Vulnerabilities in IT security pose a financial or reputational risk, whereas risks associated with IoT security are much graver as they can impact human life and well-being. In addition, chinks in IoT security posture can impact the availability of essential services and critical infrastructure for power, water supply, healthcare, transportation, and so on. This provides sufficient motivation for threat actors to plan for a sustained attack strategy.

The impact of the IoT vulnerability can be gauged from the use case described in *Chapter 5*, in the *Monitoring the condition of perishable goods* section, where a threat actor can play with the ambient conditions of the goods being transported by setting arbitrary thermostat values. Depending on the nature of the goods, even slight variations in ambient conditions (e.g., temperature changes) even for a short duration can render the complete shipment useless (transportation of vaccines is highly sensitive and temperature changes beyond the allowed thresholds for even an hour can irreversibly damage the shipment).

- Field devices are procured from third parties who may be unaware (or choose to ignore guidelines to keep costs lower) of the **secure development practices**. Also, field devices need to go through multiple stages in the supply chain (manufacturing to deployment) and may get compromised at any stage. Adding further complexity is the fact that IoT-specific security standards (e.g., EN303645) were developed quite recently and are yet to be fully enforced.

The key differences between IoT security and general/IT security discussed in this section are summarized in the following table:

IoT Security	IT Security
Attacks result in physical/material loss	Attacks result in financial and personal data loss
Hardware infrastructure (field devices) deployed in hostile/unprotected areas	Hardware infrastructure deployed within well-guarded/protected areas
Solutions managed by people with a limited understanding of security practices	Solutions are generally managed by IT experts with in-depth knowledge of cybersecurity concepts
Oftentimes, field devices are constrained in terms of processing/storage capabilities, making it difficult to execute complex security algorithms and/or manage certificates	Hardware with strong processing and storage capabilities
Diversity in field devices adding to the complexity	Limited variety of physical hardware used
Security practices are still evolving/emerging	Well-honed/mature security practices
Long hardware refresh cycle along with legacy devices	Shorter hardware refresh/upgrade cycle with no legacy devices
Difficult to provide physical security to all field devices	Easy to provide physical/perimeter security, less chance of theft

Table 11.1 – Differences between IT and IoT security

The attack surface of IoT solutions is both broader and deeper – deeper in the sense that each layer of the IoT stack can act as a potential attack surface, as shown in the following figure:

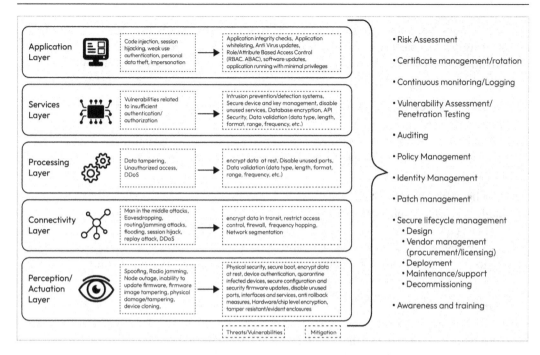

Figure 11.2 – Security vulnerabilities at different layers of IoT stack and possible mitigation

Challenges in securing IoT solutions

No one can deny the fact that for all IoT solutions, security should be considered a foundational requirement. However, some practical challenges result in suboptimal security implementations:

- **Market forces demanding faster releases and the cost of adding security features**: IoT solution architects/designers need to balance the release of a *fully secure* solution with a delayed product release. Oftentimes, this results in *good enough* security implementation but with timelines that are acceptable to the marketing/product team.

  However, with regulations/laws being formulated (and enforced), the general view is that the release of such a secure solution that is merely *good enough* won't be a possibility in the future. It is also important that internal stakeholders are aligned on the view that security is not a negotiable feature but a foundational service, without which a product may not reach the desired market acceptance. Releasing a product without this foundational service may allow threat actors to reverse-engineer the product's features, resulting in the erosion of market share, thereby putting the rationale of faster release cycles under question.

  There is a related perception that releasing a secure product results in an increase in development costs. This is a pure myth as any incremental increase in development cost is offset (many times over) by the broader market reach and longer shelf life. Additionally, most of the security libraries are free/open source and this further proves that adding security results in minimal incremental costs.

Solution providers and solution users often differ on who should bear the cost of adding security. Users believe that security is inherently a solution's property and hence should be provided at no extra cost, whereas solution providers feel that users should be willing to pay extra for a secure product.

- **Lack of awareness of IoT security**: IoT security is a relatively new topic and still evolving. As a result, there is a dearth of IoT security experts who can review/audit an IoT solution from end to end and help identify potential vulnerabilities. A person(s) operating IoT solutions may have domain understanding but isn't generally aware of the practices that are needed to protect the solution from security attacks.

  Another related issue is the unavailability of documentation for legacy devices, which makes it harder to design/implement a strong end-to-end security posture. The availability of hacking tool kits on the dark web makes the launching of security attacks even simpler. Announcing bounty programs (e.g., cash rewards) to the person(s) identifying security weaknesses is one strategy that is being effectively used by manufacturers/solution providers to tackle the challenge of attacks that are continuously increasing in stealth, complexity, and impact.

- **Non-existent or difficult-to-understand standards/laws**: Standards are almost non-existent in the IoT space and as a result, it is difficult to formulate strict policies regarding security requirements/expectations. Thus, the solution providers/manufacturers are free to implement their version of security. Another related issue is that regulations are fragmented, whereby each country is defining its own standards without any consideration for uniform standards. Moreover, although standards exist (refer to the section of this chapter on applicable standards), strict enforcement is still lacking. Like IoT standards, laws for enforcing security in the IoT space are evolving and far from being comprehensive. Also, these laws are enforceable only to vendors who supply solutions to government agencies.

  Laws should be formulated so that there is very little ambiguity on the accountability if a security breach happens – *was the breach due to the manufacturer not providing security features or was the user at fault for not using strong credentials*? Till strict laws are formulated and enforced, self-regulation by following security best practices is the most viable and practical solution.

- **Scale and diversity of IoT deployments and lack of physical security**: These factors make it difficult to provide end-to-end security. IoT devices have diverse hardware/software capabilities, functionalities, communication protocols, proprietary operating systems, and security requirements. This makes it difficult to have a consistent/uniform security strategy and implementation for all field devices. Additionally, the processing capabilities of the IoT field devices are limited and it is not suitable to execute complex security algorithms.

  Field devices are deployed in remote locations with minimal perimeter security, which provides the opportunity to carry on their malicious activities for a long time before being detected. This contrasts with a normal IT scenario where the loss of a server would immediately be reported. Additionally, the field devices are in operation for many years (sometimes even after a vendor has stopped giving support and security updates). These conditions put field devices in a precarious situation with regard to security threats.

Now that we understand the challenges that are inherent in securing IoT solutions, let us understand the security vulnerabilities that are peculiar to IoT solutions.

IoT security vulnerabilities

In this section, we will list some of the risks/vulnerabilities inherent in IoT deployments:

- **Weak, easily guessable, and hardcoded passwords**: This allows malicious/threat actors to gain unauthorized access to IoT data and systems. Similarly, the usage of default configurations is another major reason for IoT systems getting compromised. Oftentimes, the configurations are related to the features that are not used in the solution (e.g., unused TCP/UDP ports or serial ports). Using encryption keys and hash algorithms with inadequate cryptographic strength is another vulnerability that is often exploited by threat actors.

- **Vulnerabilities related to connectivity services**: Sometimes, the implementation of connectivity services is done without considering security measures (such as data encryption and secure connectivity protocols), resulting in a vulnerability that can be intercepted by threat actors to gain unauthorized access to information such as device credentials and payment information.

- **Usage of third-party libraries and/or software/hardware components**: This is another common source of security vulnerability. As an IoT solution is assembled using components from multiple hardware and software vendors/partners, there exists an inherent vulnerability as providing secure component/library vendors might not be a vendor's top priority. IoT solutions also have a higher dependency on open source libraries than traditional IT solutions, and oftentimes hackers use these libraries as testing grounds for introducing malicious logic. A good example is a critical vulnerability that was found in an open source library called Log4j that was used for logging purposes in almost all applications. The vulnerability, which was found in December 2021, put most internet-facing applications at grave risk.

  The usage of deprecated versions of third-party libraries is another factor that makes the solution vulnerable to security attacks. Sometimes, the solution provider is a source of introducing vulnerabilities by not updating the solution with the latest releases of software/hardware components.

  Having a legal contract detailing the explicit responsibility of each vendor can help in mitigating the issue – the contract should fix the responsibility of providing security updates, providing regular reports on security (vulnerability assessment/penetration testing) testing.

- **Inability to update the firmware in a secure and timely manner**: Firmware needs to be updated regularly in field devices to provide new features and/or fix known bugs/vulnerabilities. However, this can introduce vulnerabilities (e.g., firmware can be compromised during the update process). Sometimes, security patches are not released by the device manufacturer on time, and in other cases, users are reluctant to update the firmware in a timely manner (often, users avoid firmware upgrades as it involves a device restart as well). Firmware is compromised by undesirable partial or complete modification of firmware, reverse engineering of the firmware image to steal confidential information or proprietary algorithms, loading firmware onto unauthorized devices, and so on.

Some devices (especially legacy devices) lack the capability to install upgrades and/or may not support connectivity capabilities to receive updates. IoT deployments differ from traditional IT solutions in the sense that most traditional IT systems can be patched by giving users advance notice regarding pending updates. However, since IoT systems typically support critical operations and provided services are expected to run 24/7, there is a tendency to postpone the application of security patches.

Firmware upgrades carry the additional challenge where devices are expected to provide consistent performance levels, whereas parallel update processes tend to degrade performance (especially in resource-constrained devices). Ensuring the update process runs at a lower priority (at a slower rate) than the main operation is one option that can be used to overcome this challenge. Devices with sporadic connectivity should be provisioned with the flexibility to pause partial updates and resume once connectivity is reestablished. Field devices are also expected to have the ability to validate received firmware and to roll back and generate alerts/notifications if the received firmware has failed integrity validation checks.

- **Insecure data transfer and storage**: Field devices should be protected against data theft as well as the theft of firmware source code. Business-critical **Intellectual Property** (**IP**) is present in the devices in the form of algorithms and edge analytics, the theft of which can threaten competitive advantage. Some examples of critical IP include the logic for determining sleep quality in a fitness tracker and the logic of verifying the house owner for keyless entry a the smart door lock. As a result, in addition to encrypting data, the source code and related configurations should also be well protected by encryption.

- **Lack of physical security of field devices**: This results in device theft with the intention of *reverse engineering* device communications. Devices in the traditional IT enterprise are well protected (within a data center, for example). However, IoT devices are deployed in the field where they are susceptible to theft and/or misuse. Threat actors use the unauthorized access gained from field devices to launch broader attacks, such as unauthorized access to a corporate network.

    Ensuring that devices are housed in tamper-evident and/or tamper-proof enclosures along with protective mounting or casing options can go a long way in ensuring the physical safety of field devices. An example of a tamper-evident enclosure is when the device has a mechanism to ensure that any unauthorized attempt to tamper with or open it leaves evidence of the attempt – such as a paper strip around a device's enclosure that tears when the enclosure is opened by brute force. An example of a tamper-proof enclosure is one that is designed in a manner to eliminate the possibility of physical tampering or opening – a device that won't open without damage to the enclosure *or* is manufactured using screws/fasteners that require specialized tools to open. Motion sensors can also be used to detect any unexpected movement, and shielding is an effective counter-measure against side-channel attacks.

- **Lack of control or awareness about the nature of devices (or systems) that are connected to the network**: Organizations don't have an accurate inventory of the devices and software systems that are connected to the network by the employees or users and hence can't enforce security standards for these devices.

Major IoT security breaches

In this section, we'll list some of the major security breaches that have occurred in the IoT domain in the recent past:

- **Stuxnet**: This attack occurred in 2010 when a virus named Stuxnet caused damage to a nuclear centrifuge in Iran. Virus-modified commands were sent to the **Programmable Logic Controller (PLC)** and targeted industrial machines running the **Supervisory Control and Data Acquisition (SCADA)** protocol.

- **Vehicle hacking**: This attack happened in 2015 when two security researchers hacked a jeep and played with the controls by changing radio channels and turning on wipers and air conditioners. This forced the jeep manufacturer to recall 1.4 million vehicles to patch the exploited vulnerability. In a similar attack, one hacker was able to demonstrate how easy it is to drain the vehicle's battery by using its **Vehicle Identification Number (VIN)**.

- **Mirai botnet**: This is one of the biggest IoT attacks to date and occurred in 2016 when it attacked Dyn (one of the largest DNS providers), resulting in the disruption of the operation of the biggest companies, such as Twitter, Amazon, and Netflix. The attackers used IoT devices such as routers and IP surveillance cameras to launch a DDoS attack.

- **St. Jude cardiac device attack**: This attack occurred in 2017 when the **Food and Drug Administration (FDA)** declared that cardiac devices (such as pacemakers) manufactured by St. Jude Medical had a vulnerability where hackers could remotely stop their operation. Similarly, in 2015, an internet-connected drug infusion pump was found to be vulnerable to remote control with a hacker being able to increase or decrease the dose level with direct risk to a patient's health.

- **IP-based video cameras**: Attackers have targeted IP-based video cameras to capture the video feed and, in some cases, tweak the capture settings. In 2021, a major attack was launched on a start-up in Silicon Valley (Verkada) where threat actors executed malicious code by gaining privileged access. This was done to launch an all-out attack on the enterprise network, thereby exposing confidential and business-critical data. Additionally, they were able to access live (and archived) video streams from 150,000-odd cameras that were managed by the start-up.

- **Connected HVAC**: This vulnerability was exploited by hackers to sneak into Target's (a major retail chain's) financial system. The hacker's modus operandi was to steal network credentials from the HVAC vendor.

After understanding the nature and impact of IoT security breaches that have occurred in the past, let us understand how following a set of best practices/guidelines in a consistent manner could have prevented the stated (and similar) breaches/attacks.

Mitigating IoT security vulnerabilities

Providing a robust security posture is a complex endeavor and requires the active involvement of different parties throughout the solution life cycle. Following the guidelines (or preferably a combination thereof) mentioned in this section would help to provide a credible defense against potential threats/vulnerabilities:

- **Using AI/ML techniques**: These are effective deterrents against increasingly sophisticated security attacks. AI/ML techniques can monitor network traffic as well as device data to flag suspicious behavior. Manual monitoring is no longer an option due to the ever-increasing list of attack vectors.

    These techniques rely on using context-aware access controls at all layers of the IoT stack to allow only the expected actions and behaviors. The system behavior is monitored for a reasonable period to establish a baseline of expected behaviors and anomaly thresholds and the accumulated baseline data is used to flag malicious and/or unauthorized activities.

    In other cases, the process involves collecting, correlating, and analyzing data from multiple IoT data sources and combining it with a threat intelligence knowledge base. Prebuilt AI/ML models then analyze the data for possible vulnerabilities and provide recommendations for fixing the identified vulnerabilities. If multiple vulnerabilities are detected, they are prioritized by these models. It is ideal for all the identified vulnerabilities to be resolved; however, this is seldom possible due to effort and cost considerations. With the increase in the maturity of AI/ML models, recommendations can also be implemented in an automated fashion.

    AI/ML techniques are effectively used in an IDS where they continuously monitor the network for suspicious traffic patterns (an alarm should be raised if a device that is designed to send data once a day starts sending data multiple times a day) and leverage data from previous attacks to predict the occurrence of a security attack and recommend steps to contain the damage.

- **Developing a toolkit of security implementations**: Solution providers can do this and then customize the toolkit to suit specific domain or device family requirements. This will overcome the complexity of IoT solutions due to the scale and diversity of IoT devices and will allow the solution provider to tailor the rigor of security implementation as per the customer's needs. Additional steps that a solution provider can take to ensure a secure product are as follows:

    - **Leveraging security best practices for solution development**: This includes assigning unique identities to all the devices/components, performing design and source code reviews (specially focusing on memory leaks, buffer overflows, backdoors, and validating or sanitizing data inputs), and using the latest operating systems, toolchains, libraries, and so on. Periodic reviews and audits of device, server, or network logs, assigned privilege levels, and configurations and maintaining a detailed incident response plan are equally important for establishing a strong security posture.

    - Unique identities are required for device management features (remote access, patch management, and so on) and help prevent unauthorized operations such as device cloning. In general, devices are provided with minimal permissions during the device manufacturing process and full identity and permissions during the field installation (refer to the *Device management architecture* pattern covered in *Chapter 2* for more details). This approach is especially useful if the intended usage of the device is not known at the manufacturing stage. The device identity is created from multiple device-specific attributes, as shown in the following figure:

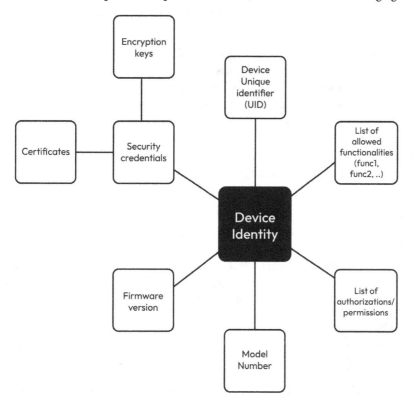

Figure 11.3 – Key attributes of device identity

- **Providing regular firmware updates for field devices and backend systems**: Users should be encouraged to update the firmware at *first use* as firmware would have been updated between the time it was manufactured and the time the product reaches the customer site. Enabling the device's auto-update firmware upgrade feature is another good practice that can be followed. An alarm should be raised if any device is skipping the upgrade requests.

- **A strong password and authentication mechanism** (such as multi-factor authentication or biometrics) should be used. A manufacturer can also set up policies that enforce strong passwords and require the user to change passwords at regular intervals. The usage of a certificate issued by a public CA that is based on asymmetric encryption makes security implementation completely scalable (by eliminating the need to have a dedicated private server for authentication purposes).

- **Rotating passwords and certificates**: Providing unique identities to each of the devices in the form of device certificates – certificates are loaded into the device during the manufacturing stage and are activated later. An important point to consider here is that not all field devices would be capable of authenticating certificates (due to limits in terms of processing power and storage capabilities) where alternate authentication mechanisms such as keys and passwords need to be used.

- **Securing critical data**: TPMs and TEEs can be used to secure critical data (e.g., encryption keys) along with disabling unused ports (e.g., debugging ports) and solution features. As most field devices are manufactured to serve very specific functions, most of the ports remain unutilized and hence should be disabled. Similarly, functionalities that are not relevant for a particular deployment should be disabled (*principle of least functionality*) and allowed functionalities should be executed at the least privilege level with access to only those resources that are required to complete the functionality. Maintaining a whitelist of devices that can connect to the network is another effective mechanism to minimize the attack surface.

- **Encrypting data at rest and in motion and network isolating the device**: This results in a reduced attack vector as well as a reduced blast radius. To avoid data misuse, it is prudent to meticulously identify/review data storage and transmission requirements as unused data results in an increased attack surface. Deleting or transferring data to a secured location after a predefined interval is another effective approach for reducing the attack area. Ensuring that field devices can connect to the internet via firewalls, proxies, and so on, along with the usage of access control lists, results in an enhanced security posture.

- **Maintaining an updated inventory of field devices**: Such an inventory would include, among other information, manufacturer details, device ID, serial number, location, hardware/software configurations, and software/firmware versions. This information can be part of the **Asset Management Database** (**AMDB**) that is already maintained by solution providers. Devices that can't support modern security mechanisms should be separately identified and placed within a separate network. Additionally, a plan to replace/retire these legacy devices in due course should be formulated.

- **Using newer technologies**: Technologies such as blockchain can be leveraged to help improve the security posture as these technologies provide secure, decentralized, and tamperproof data storage and transmission. Similarly, it is prudent to avoid using legacy (non-secure) protocols such as Modbus and the usage of protocol converters should be explored to convert unsecure protocols to secure protocols, such as HTTPS, TLS, and **Secure File Transfer Protocol** (**SFTP**).

- **Monitoring vulnerabilities**: It is important to ensure that the constant monitoring of vulnerabilities reported for the devices in the public domain and timely patching are done. This relates to the *Device management* architectural pattern explained earlier in *Chapter 2*, which is used to manage the life cycle of the field devices. The security recommendations need to be followed for all life cycle stages (registration, activation, commissioning, de-provisioning, and patch management). Specifically, at the time of de-commissioning, all the sensitive data needs to be deleted.

- **Anonymizing data**: Before storing data on the DG or the central server for long-term processing, care should be taken that personal/sensitive data is adequately anonymized and/or redacted. Implementing security logic (network segmentation, encryption/decryption, certificate rotation, and maintaining device whitelists) on the DG is generally preferred, especially in cases where devices attached to the DG can't implement these mechanisms.

- **Logging of operational and security events**: Components of the IoT system should support this feature and logs needs to be analyzed at periodic intervals to detect/respond to security incidents in a timely manner. These logs are also useful for doing post-facto analysis of a security incident. Wherever possible, the system should be configured to not perform any operation in case event logging is not possible. Often, the mentioned log analysis is done by a dedicated system (SIEM) outside of the main IoT data and operational flows.

- **Implementing standard guidelines and rating mechanisms**: Doing this (preferably through a trusted third party) will help consumers to easily understand the level of security implementation. This will not require consumers to understand the security nuances in detail, but they will still be able to gauge the suitability of the product as per their risk appetite. This would be very similar to the ratings used by automobile manufacturers to indicate the level of pollution emitted by a particular vehicle. In fact, these ratings would indicate the rigor of security testing done by the manufacturer and would involve measures such as penetration testing, vulnerability assessments, and security compliance/certifications.

- **Maintaining a security-first mindset**: Such a mindset, where security is considered a crucial part of the development and release processes and not as an activity that must be *bolted on* as part of the final release (also referred to as the **Secure Development Lifecycle (SDL)** where security measures are embedded in each phase of solution development life cycle), is crucial. This starts with security assessment/threat modeling, which identifies risks, gaps, and vulnerabilities in field devices and communication channels. Security assessment helps to identify actors who stand to gain by exploiting the vulnerabilities, their motives, and the sophistication of tools/tactics that they may use. The assessment also helps to identify the high-sensitivity/high-impact components so that security mitigation efforts can be suitably prioritized.

  Based on the identified risks, organizations can then formulate an appropriate security model and related policies. After the security model and policies are defined, the next step is to define the security architecture. The security architecture helps to translate abstract security needs (as expressed by the security model and policies) to actual mitigation procedures. *What type of anti-tampering mechanism should be employed to prevent device misuse? What level of security is required for the device and central server?* As can be seen, these mitigation steps need to be developed for all the assets that were identified as part of the assessment exercise.

- **Providing network segmentation for IoT devices**: Network segmentation is done to enable granular control over data traffic between devices and the central server. This is done to reduce the blast radius in case the threat actor infiltrates the network. Network segmentation is provided by using **virtual local area network (VLAN)** configurations or firewalls. Another possibility is to maintain *islands* of small networks within a single network that communicate only with each other. Segmentation is done based on the connectivity and security requirements of each field device (internal connectivity, connectivity to one destination, or public network connectivity). Firewalls can also be configured with specific rules to accept connection/data requests from specific device IP addresses or geographical regions only.

Creating and maintaining an updated version of the network architecture showing how field devices are connected to each other and the device gateway/central server is a good practice that provides multiple benefits. This architecture is generally part of the playbook that lists the known security incidents and their mitigations.

- **Responsibility segregation**: Clearly segregating the responsibilities of the solution provider/ manufacturer and solution user helps to clarify the activities each party has to perform to mitigate security risks. In general, the user needs to shoulder responsibility if the breach is due to wrong usage (or inadequate adherence to precautions) and the manufacturer is liable if the issue is due to wrong design, manufacturing, or inadequate security validation/verification. The manufacturer is also required to provide a security guidelines/best practices document (normally a part of the solution instruction manual) to avoid the inadvertent exposure of the product to security threats and vulnerabilities. Similarly, companies that host data on behalf of their customers would be held accountable if data is lost or exposed.

Possible approaches for mitigating security vulnerabilities discussed in this section are summarized in the following figure:

Figure 11.4 – Approaches for mitigating security vulnerabilities

Domain-specific security considerations

The data sensitivity and hardware or software capabilities of both the backend system and field devices may vary from one domain to another (connector consumer devices such as smart fridges typically run on constrained hardware, whereas the autonomous car would typically have powerful processing capabilities). The importance attached to the three pillars of security (CIA) varies from one domain to another or one use case to another – the video feed from a camera installed in a public place (traffic intersection) might not be confidential but the data feed from a person's living room would be confidential. Although the fundamental principles of providing IoT security are similar across domains, there are certain implementation nuances that you should be aware of, which are described in this section:

- **Difference in data sensitivity across different domains**: The volume and level of sensitivity of the transmitted data varies from domain to domain and accordingly, the rigor of the security implementation will vary from domain to domain – consumer devices (such as a smart thermostat) will have a different set of security requirements than medical devices (connected pacemaker). Accordingly, the laws governing data security/privacy will vary from one domain to another.

- **Difference in profile and objectives of threat actors**: Consumer devices (such as thermostats, smart speakers, smoke alarms, and doorbells) are less prone to organized and/or state-sponsored attacks. However, they can be targeted for ransomware attacks (such as a doorbell ringing at odd times unless the user concedes to threat actor demands).

Due to the differences in the potential impact of vulnerabilities, security approaches vary across domains, for example, privacy is more important in the consumer than the agriculture and manufacturing domains. To illustrate this more clearly, we'll discuss the security challenges for two domains in the following list:

- **Manufacturing**: This domain has challenges of legacy machines with long life cycles (average life cycle time of 10 to 30 years) with a mix of brownfield and greenfield deployments. Robust security implementation is almost a mandate in the manufacturing domain as attackers use vulnerabilities in the OT infrastructure (where protection is generally weak) to launch an attack on the IT infrastructure. This increased attack surface can have grave consequences and result in financial loss and/or pose a threat to workers' lives.

- **Consumer**: There have been quite a few security incidents (mainly privacy related) with the prime reason being that users are not aware of the potential risks of not updating firmware and security patches on time. Users tend to value simplicity and ease of usage much more and expect security concerns to be handled by the manufacturer without their explicit involvement. Also, users tend to replace/upgrade their devices quite often (the average life cycle is 2 to 5 years), which provides an opportunity for manufacturers to pass on the latest security implementations with each new product iteration. As users in the consumer market are always interested in the latest and newest offerings, the solution provider needs to deliver in shorter cycle times and security becomes an unintended casualty.

Although domain-specific applications can be exploited by attackers, data from applications from different domains can be combined to misuse the information at a larger scale. In fact, data collated by different domains is different and can be combined to create a complete profile of a user and their behavioral characteristics. As an example, IoT applications in consumer space can determine a person's habits and behavioral patterns, whereas autonomous vehicles can provide details regarding the routes preferred by the user. Data gathered by retail domain applications can be further exploited to gain knowledge regarding product/brand preferences. Similarly, IoT applications in the healthcare domain can provide information about illnesses, treatments, and so on. The accumulation of such holistic data by threat actors can be quite lethal.

We've learned about domain-specific considerations for designing secure IoT solutions. The knowledge gained in this chapter can be further supplemented by going through the security standards that are listed in the next section.

Applicable security standards and best practices

As mentioned earlier in the chapter, there are a few standards that specifically cover IoT security-related compliance requirements. Here are some references to some of the prominent standards and best practices:

- The *ETSI EN 303 645 V2.1.1* standard: `https://www.etsi.org/deliver/etsi_en/3` `03600_303699/303645/02.01.01_60/en_303645v020101p.pdf`

- *NIST Cybersecurity for IoT Programs*: `https://www.nist.gov/itl/applied-` `cybersecurity/nist-cybersecurity-iot-program`

- *IoT Security Best Practics*: `https://www.iotsecurityfoundation.org/`

- Best practices for threat modeling: `https://owasp.org/www-community/Threat_` `Modeling_Process`

- *IoT cyber security regulations across the world*: `https://cetome.com/panorama`

Summary

This chapter provided insights into why security is important in IoT solutions and how the implementation of security measures is different in IoT compared with general IT solutions. Also, it is prudent to leverage the existing knowledge base of generic IT security and tailor/enhance that to suit the needs of IoT security. Some IoT vulnerabilities that were exploited by threat actors in the recent past were also discussed to give a perspective of how vulnerabilities differ in IT vis-à-vis the IoT space.

One key takeaway from this chapter is that there is no single solution that can be used to mitigate IoT security risks and a combination of technical, operational, and organizational measures can help in mitigating potential vulnerabilities. IoT security can be best accomplished if the mitigation strategies use a combination of both the *defense in-depth* (analyze possible risks and their mitigations at all layers of the IoT stack) and *defense in-breadth* (consider multiple mitigation strategies for each layer) approaches.

Protecting IoT solutions is a continuous journey as a solution that is fully secure may become unsecure tomorrow due to the increased usage of sophisticated tools and tactics. Although there is no such thing as *foolproof security*, what is required is to provide security against known risks and perform regular threat modeling/audits so that it becomes increasingly difficult for threat actors to exploit vulnerabilities that exist in the system. Equally important is to have an incident/response plan ready to be executed in case any security vulnerability/threat is detected. IoT architects are expected to balance the requirement of releasing a fully secure solution with a delayed product release or release it with *good enough* security with timelines that are acceptable to the marketing/product team.

As with any other technology, the field of IoT security is constantly evolving. With increased sophistication provided by AI/ML algorithms, future IoT systems will be able to not only detect security gaps but also execute require mitigation steps with no (or minimal) impact on system operations, resulting in truly *self-healing* systems.

So far, we have focused on architecting and implementing IoT solutions. However, in the next chapter, we will be taking a slight diversion to explore how emerging technologies, such as Web 3.0, 3D printing, 5G, and social media, can augment IoT for the development of more compelling solutions.

Part 4:
Extending IoT Solutions

This part provides details about how IoT solutions fit into the overall ecosystem – for example, how combining IoT with related technologies such as blockchain and AR/VR can result in innovative and rich use cases. The last chapter of the book focuses on providing guidance on how to handle the practical challenges encountered when building IoT solutions.

This part comprises the following chapters:

- *Chapter 12, Exploring Synergies with Emerging Technologies*
- *Chapter 13, Epilogue*

12

Exploring Synergies with Emerging Technologies

In the previous chapters, we saw that IoT technology helps merge real and virtual worlds, where data from the real world is aggregated or processed at a central location, and generated insights are used to trigger actions back in the real world. This capability allows us to develop very rich, diverse, and useful applications. However, this capability is enhanced multifold if we combine IoT with other related technologies (such as blockchain, metaverse, generative AI, and others) – the sum is more than its parts.

The chapter covers the main technologies that can effectively complement IoT, but the list of technologies is not exhaustive; listing all the technologies is not practically possible as new technologies are continuously emerging. Additionally, this chapter intends to provide you with a glimpse of the possibilities that exist when we combine IoT with other technologies (*art of the possible*). However, like any other technology, the technologies listed in this chapter may have inherent limitations, vulnerabilities, and/or risks associated with them (as an example, AI/ML technology may suffer from different biases and there is a concern regarding large language models being used from a privacy standpoint). Hence, it is recommended that you study the pros and cons associated with a specific technology in detail before you plan to build the required application or use case.

The main objective of this chapter is to introduce you to additional technologies that can complement IoT capabilities and will help solve problems that can't be solved by leveraging IoT technology alone.

In this chapter, we will provide an overview of various technologies, explore the benefits of combining those technologies with IoT, and finally discuss some concrete examples of how a combination of technologies can help solve real-world problems. For clarity, the technologies mentioned in the chapter relate to different layers of the IoT reference architecture's layers, as is shown in the following figure:

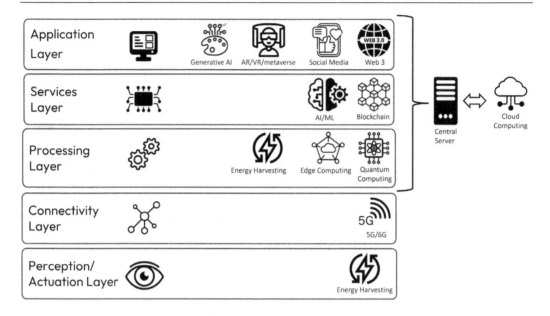

Figure 12.1 – Emerging technologies mapped to the IoT reference architecture

Let's start with a brief overview of blockchain.

Blockchain

Blockchain is a decentralized digital ledger that stores all transactions in a distributed fashion (that is, each node in the network gets a transaction copy). Every transaction in the blockchain (that is, every block) is stored alongside a timestamp that is validated and approved by all the entities or nodes in the network. Once approved, the transaction is aggregated as a block and replicated on each node in the network. As a result, transactions maintained by blockchain technology are considered immutable and tamperproof. These capabilities enable blockchain to solve one challenge – that is, integrity. Here, there are three key security challenges that any IoT solution is expected to mitigate, which form the **Confidentiality, Integrity, and Availability** (**CIA**) triad. In *Figure 12.1*, blockchain is represented using this image:

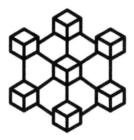

Figure 12.2 – Blockchain

Benefits of combining IoT and blockchain

As we saw in *Chapter 11*, field devices aren't capable of implementing strong security or cryptographic algorithms. Blockchain helps provide robust and scalable security because of the following reasons:

- Blockchain's decentralized nature eliminates the possibility of a system being compromised due to a single security vulnerability point.

- The participation of field devices in blockchain transactions without the involvement of a central server improves overall fault tolerance by eliminating a single point of failure.

- Leveraging blockchain makes the solution horizontally scalable as one entity (that is, a central server) isn't loaded with extra processing when a new field device (a node) is added to the network – security-related processing (encryption, authentication, and so on) gets distributed evenly among all the existing nodes.

- Blockchain provides strong integrity (eliminating the possibility of data being tampered with while in transit or at rest) as all the events (data transfer, device states, and others) are maintained in an immutable ledger – it is almost impossible for the threat actors to cause post facto modification of data.

- The integration between IoT and blockchain would result in the development of efficient and secure payment transactions. Most of the current payment mechanisms (credit cards and so on) are reasonably secure, but transactions that are enabled by blockchain are point-to-point and don't involve intermediaries (for example, banks).

> Important note
>
> Integrating blockchain with IoT comes at a cost as it involves extra processing (and therefore higher power consumption) by field devices, and it also consumes extra bandwidth to support additional data communications. Also, extra storage capacity is needed by field devices as the blockchain ledger needs to be replicated at each device.

Possible use cases

Blockchain technology enhances the security posture of IoT use cases. Accordingly, it can be applied to any of the use cases detailed in *Chapters 4* to *8* to provide security in a decentralized manner.

As an example, adding blockchain security to a condition monitoring use case (detailed in *Chapter 5*) would reduce the possibility of **man in the middle** (**MITM**) attacks and eliminate the possibility of tampering with inflight sensor data due to strong authentication and integrity safeguards built into blockchain technology. This would also ensure the shipment's accurate traceability during its entire journey from producer to consumer and enable the consumer to reject the shipment if there is any indication of data tampering or if the desired storage conditions were not maintained at any point during transit.

Generative AI

Generative AI has received considerable attention in the recent past. It is a form of AI that can generate unique content (images, video, audio, tests, and even code or logic) based on the provided keywords, where the content's quality closely mimics the content a human would have created.

This aspect of generating unique content differentiates generative AI from traditional AI, which relied primarily on a predefined set of patterns, rules, formulas, and more. Additionally, generative AI is trained on a considerable volume and variety of input data/content (images, sound, video, programming patterns, and more) and it can *self-learn* or *adapt* to insert novel/unique elements into the generated content rather than simply combine content from multiple input sources. In addition to being trained on a huge corpus of content, generative AI leverages **deep learning** techniques to continuously fine-tune parameters, resulting in more and more authentic output with time.

We represent it in *Figure 12.1* using the following image:

Figure 12.3 – Generative AI

Some popular current (that is, in 2023) examples of generative AI implementations are listed here:

- **ChatGPT/Google Bard**: A **natural language processing** (**NLP**) chatbot that can understand complex queries and generates human-like responses.
- **DALL-E/Midjourney**: The user provides a detailed description of the type of image required via prompts. Then, these tools generate realistic images.
- **resemble.ai**: Generates audio samples.
- **Amazon CodeWhisperer/Copilot**: A tool for generating source code fragments.
- **Adobe Firefly**: A tool for intelligent image editing and content/image creation.

Now that you understand what generative AI is, let's look at the benefits of integrating it with IoT.

Benefits of combining IoT and generative AI

IoT augmented with generative AI capabilities can help create intelligent and creative systems where they can interpret natural languages and generate responses that are indistinguishable from human responses. However, this requires significant investment in terms of model training, retraining, and finetuning to get the desired results. Normally, a base or generic model is taken as a starting point and is further fine-tuned for a specific use case or problem domain.

A good example is the healthcare domain, where structured, semi-structured, and unstructured data related to patients, symptoms, and clinical tests (vital parameters, MRIs, X-Rays, CT scans, and so on) is analyzed to predict disease onset and provide tailored diagnostics recommendations. Additionally, the information related to treatment (precautionary measures and medicine doses) can be communicated in a personalized manner considering individual preferences and acceptability.

Combining IoT with generative AI has the potential to drastically improve the interaction between IoT systems and humans. At the time of writing, most of the commands are taken either from mobile phones or HMI (as mentioned in *Chapter 10*) and output is displayed on screen or sounded on a speaker. Generative AI can help us understand the commands that are passed in natural language and generate more human-like voice responses (for example, "Which of the rooms in a building have temperatures above 25^0 C?").

Possible use cases

Generative AI can be used to design 3D enclosures for IoT field devices without the need to specify intricate design details. The design details can be provided in plain English and generative AI can produce detailed instructions that can directly be fed into a 3D/4D printer. A combination of IoT, generative AI, and 3D printing can be used to create hyper-personalized products where product features and their design (that is, the look and feel of the enclosures) are customized to suit an individual's tastes.

Generative AI can be used to design visual interfaces (dashboards) for consuming IoT data/insights. Similarly, the technology can be used to customize dashboards that have been created for one customer so that they suit the requirements of other customers (change in logos, relative placement of trend charts, and other visual elements). Generative AI technology can help generate automated reports by anonymizing private data so that it complies with regulatory norms.

In more advanced applications, generative AI can be used to dynamically alter the processing flow (or workflow) based on the observed sensor data – this would involve further training of the generative AI model. Consider a scenario where static event/action rules executed by LRE or GRE are replaced with dynamically generated rules (rules that are created on the fly based on real-time data, as well as the overall spatial and temporal context).

Large language models

Large language models (**LLMs**) are extensions of deep neural networks that are trained on large datasets and tuned for a vast number of parameters (billions of parameters) and are capable of understanding and generating human-like content. We represent it as follows:

Figure 12.4 – LLMs

Although LLMs can understand other media types as well (such as audio, video, and images), their focus is primarily on understanding textual input and generating textual output. The key difference between LLMs and other query resolvers/chatbots is that LLMs can build an understanding of the overall context by correlating a set of queries (and responses) and can answer complex queries in a more nuanced manner. This enables it to perform tasks such as **computer vision** (**CV**) and NLP with a precision that is close to human-generated output.

Training over a large dataset enables these technologies to generate natural (human-like) responses rather than *canned* or *superficial* responses (for example, these models can consume a large chunk of text and provide a summarization that isn't a simple repetition of selected portions of the main text but closely mimics the way a human would articulate the summary). LLMs can be put to other uses, such as scanning an image and providing a text summary and/or providing a relevant caption/title. This technology provides capabilities such as sentiment analysis, more natural and nuanced language translations, and attribution or causal analysis, and helps generate creative and unique content on demand. These capabilities equip the technology to be used for text translation, text summarization, generating congruent groups of sentences (for example, generating a complete story by inputting a partial story), natural query responses, and more.

Both generative AI and LLMs are based on deep learning techniques and leverage foundational models as a core technology. However, generative AI is a more generic term and LLM can be considered a specific implementation of generative AI, where the focus is primarily on understanding and generating textual content.

Benefits of combining IoT and LLM

IoT can be combined with LLM to analyze user feedback on social media regarding product quality. These comments can be analyzed and data from the IoT sensor data from the product can be used to verify the feedback's veracity (whether the user providing feedback has used the product or not); then, the authentic feedback can be used to generate future product enhancements. These enhancements and recommendations would be like the requirement analysis or user stories documentation that is currently prepared by product managers.

Possible use cases

Since LLMs can interpret images/videos, this technology can be used to interpret IoT data visualizations (for example, dashboards) in the absence of the equipment or machine operator. For example, if any anomaly is detected in an assembly line, an LLM can work in conjunction with generative AI to generate a detailed alarm message and send the same to the operators or workers on the shop floor (using speakers or flashing it on the available video terminals), along with specific instructions to resolve the anomaly. Operator and LLM implementation can work in tandem in monitoring the assembly line situation – with LLM providing routine monitoring and the human operator applying their intuition on top of the LLM-generated data or insights to identify complex patterns.

LLMs can also be used to provide automated (and more natural) technical support for operational and diagnostics issues reported by IoT device owners. In a more practical scenario, the LLM would already exist for generic use cases and would need to be customized/fine-tuned to fulfill the needs of IoT use cases – refer to the *Transfer learning* section of *Chapter 10*.

AI/ML

AI/ML is a general term that encompasses a lot of other technologies, such as deep learning, LLMs, and generative AI. Here, it is represented as follows:

Figure 12.5 – AI/ML

AI/ML integration was described in detail as part of the *AI/ML integration* pattern in *Chapter 3*.

Benefits of combining IoT and AI/ML

One benefit of combining AI/ML with IoT is that you can ingest and process high-volume and velocity data produced by IoT sensors in almost real time to derive timely/actionable insights. In addition to efficiency and performance gains, integrating AI/ML with IoT will free decision-making from biases inherent in human thinking. As a result, the combination of these technologies would result in more efficient, timely, and accurate analysis, which is a major advancement over traditional static rules (*if X, do Y*) or algorithms.

AI/ML integration helps in improving the quality and accuracy of the raw data obtained from sensors by removing inconsistent data (such as outliers) and/or imputing or inserting missing data. The role of AI/ML in improving the security posture by analyzing the possible threats and implementing automated responses was covered in *Chapter 11* – performing penetration testing and vulnerability assessments in an automated manner is another example of using AI/ML for enhancing the security posture of IoT solutions.

AI/ML can help in correlating the data from different sensors with external data sources (for example, weather station) to generate more relevant or holistic insights (a sudden increase in a heartbeat but without a corresponding increase in motion may indicate the possibility of a heart attack, for example). AI/ML is also a key enabler of sensor fusion (as mentioned in *Chapter 10*). The role of AI/ML in enabling more complex simulations (*what-if* scenarios) in the context of **digital twins** is also worth noting.

AI/ML helps generate and discern a product's usage analytics (product features that are frequently used vis-à-vis the features that are rarely used) and that's a vital input for future product revisions or to tailor products based on customer demographics, geographic locations, and more.

Possible use cases

One interesting use case is to dynamically determine whether a particular data processing requirement should be performed locally (that is, at the edge) or at a central server. This would involve making drastic improvements from the current implementations where there is no such flexibility and processing data or rules in either the cloud or the edge in a fixed and predefined manner.

Other general uses of AI/ML in IoT use cases such as data quality improvement, automated security monitoring, and others were also described in previous chapters. Essentially, AI/ML enables automated decision-making, which can be leveraged in multiple use cases and diverse domains, as illustrated in the following figure:

Smart City	Analyzing video/audio feeds from cameras or sensors to detect suspicious behavior, such as fighting, gunshots, and theft
	Using autonomous vehicles
	Monitoring traffic at intersection points to control traffic lights in a dynamic and adaptive manner
Assisted Living	Analyzing data from video cameras or sensors, such as fall or panic sensors, to determine the fall possibility
	Monitoring sugar levels and generating alerts if anomalous readings are detected
Retail	Allocating staff dynamically in shops based on the footfall data to minimize checkout times
	Mall owners charging rentals based on the actual footfall
	Tracking and optimizing shoppers' journeys
Manufacturing	Assisting in automated quality control by separating faulty parts and implementing predictive maintenance
	Providing recommendations for optimizing assembly line operation
Energy/Utilities	Predicting demand and supply

Figure 12.6 – Use cases enabled by IoT and AI/ML integration

Immersive technologies

Immersive technologies, including **augmented reality** (**AR**), **virtual reality** (**VR**), and the **metaverse**, provide an immersive and engaging mechanism for consuming IoT data and insights. Here, it is represented using the following image:

Figure 12.7 – Immersive technologies

Traditionally, content or data is consumed using desktop or mobile screens, but these immersive technologies offer an enriching experience to the user, as described here:

- **AR**: As mentioned in *Chapter 10*, AR helps overlay IoT content (analytics results, sensor operational or metadata, and so on) with the user's current view – that is, AR allows you to augment the real-time view of the user with context-sensitive IoT data and information. To consume AR data, an additional device (wearable glasses, a mobile device, and so on) is required that superimposes IoT data on the real-time video feed obtained from the device's inbuilt camera. As an example, a person can see the house plan along with the actual version for faster troubleshooting, as shown in the following figure:

Figure 12.8 – AR blending real and virtual content

- **VR**: VR differs from AR because, in VR, the view is completely replaced with alternate content (or alternate/different reality) via a headset. VR also provides a higher immersion level compared to AR by replacing the complete view. Also, content changes as per the direction of the headset, giving the user the illusion that they are present at the place of action. Typical use cases of VR include training, entertainment, virtual sightseeing, and remote meetings.

- **Metaverse**: The metaverse is an extension of VR where an alternate world is created that is inhabited by virtual identities (avatars) of the inhabitants, along with related material objects and processes. It is predicted that a person would be able to perform most of their daily activities, such as work, shopping, socializing, and leisure/play in the metaverse.

  The main difference between VR and the metaverse is that of **scale** and **complexity**. Considering a manufacturing plant as an example, VR would cover one or more assembly lines, whereas the metaverse would cover the complete plant and may even include upstream and downstream entities or processes involved in the supply chain. Another difference between VR and the metaverse is that VR loses its current state at the end of the session, whereas the metaverse maintains the state or environment while users enter and exit the metaverse at their convenience.

Now, let's take a look at the benefits of combining these technologies with IoT.

Benefits of combining IoT with immersive technologies

AR, VR, and the metaverse primarily differ in the level of immersivity provided. Accordingly, these technologies can be used in IoT use cases to present the IoT output (data, analytics, and so on) with varied levels of immersivity. IoT can be considered as an agent that keeps the data synchronized between real and virtual worlds (the world of AR, VR, or the metaverse).

The relationship between IoT and immersive technologies can be understood using the following figure, which shows one of the possible deployment options:

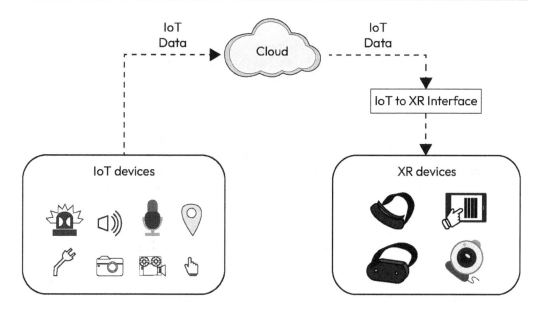

Figure 12.9 – IoT acting as a bridge between real and virtual worlds

Possible use cases

The need to consume information in an immersive manner is common in different domains (industrial, manufacturing, retail, consumer, and others). IoT can be combined with AR/VR/the metaverse to cater to the requirements of different domains, as explained here:

- The digital twin implementations of architectural patterns (detailed in *Chapter 2*) can add AR/ VR/the metaverse as a visualization mechanism. As described, data from the field devices needs to be constantly fed into the digital twin so that it can represent the true state of the actual real system. Similarly, actions initiated on the digital twin by the user need to be implemented (actuated) in the real world. IoT enables these two aspects (sensing and actuation), whereas the state of the digital twin (and by extension, the real world) can be effectively displayed/ visualized using AR, VR, or the metaverse.

  As an example, avatars in the metaverse can more closely mimic reality if their attributes (facial expressions, bodily movements, and so on) are captured alongside other parameters. Likewise, elevated heart and breath rates from a person in the real world can be used to show the metaverse avatar in a fatigued state. Some of the static attributes such as a person's height, weight, or the color of their eyes can be configured as avatar metadata, but dynamic attributes (heart rate, blood pressure, and more) need to be captured in real time by IoT sensors for more holistic rendering. In generic terms, IoT sensor data provides contextual and situational awareness of the virtual world and takes sensory inputs (triggers in the form of button presses, body or head movement, hand gestures, and so on) and replicate the same in the real world.

> **Important note**
>
> To effectively simulate the real world, immersive technologies need to combine data from multiple sources (field data, person data, haptic data from headsets, and so on) and present it in a unified view. Sensor fusion (outlined in *Chapter 10*) plays a vital role here.

- Immersive technologies help you experience the real world from a remote location. As an example, consider a metaverse scenario where automobiles such as cars are exhibited and users can experience the aesthetics of the car remotely and have the almost real experience of driving the car. For this to work, IoT and immersive technologies need to work in conjunction to synchronize data between real and virtual worlds – users would receive sound, tactile, and other sensory experiences (driving on a bumpy or smooth road) and would also be able to control the car's functions.

- Immersive technologies also help in remotely monitoring and controlling a machine or piece of equipment. Technicians can get a real-time view and feel (vibration, thermal behavior, and so on) of the equipment's or machine's state and initiate troubleshooting steps remotely. Similarly, supplementing immersive technologies with ambient data gathered by IoT sensors can make virtual events, visits to shopping malls, or conferences realistic, livelier, and wholesome.

- AR is specifically suited to maintenance activities as the internal structure/mechanisms of the machine can be known even without opening the machine's enclosure. IoT sensors can provide situational or locational awareness that can help you present the data at the required granularity level – it can provide a summarized view in case a person is far from a machine and provide more detailed information as the person moves closer to the machine. Also, a person at the remote site can communicate their observations to the technical expert stationed at the main office and seek their guidance or opinion regarding the correct maintenance steps.

  Visual cues provided by AR technology can also be used to locate a misplaced part of a machine or equipment where IoT would supply locational information (using sensing technologies such as GPS, RFID, and so on) and AR would help guide the worker to the relevant location in the factory.

  AR and IoT can also augment human capabilities, enabling people to perform their tasks in an efficient and foolproof manner – visually inspecting finished parts can be done by the worker while the image of the correct product is projected using an AR headset, which results in them not needing to switch their attention between the product and the parts on the assembly line.

Next, let's take a look at the role that IoT can potentially play in 3D and 4D printing.

3D and 4D printing

3D/4D printing creates customized parts by depositing layers of molten substance. We represent it as follows:

Figure 12.10 – 3D printing

Before understanding how this innovative technology can augment IoT use cases, let's understand these two technologies a bit more:

- **3D printing/additive manufacturing**: 3D printing involves creating a 3D object by adding/depositing raw material in a layer-by-layer manner – 2D raw material is layered one over the other, thereby creating a third dimension. This contrasts with traditional manufacturing, where products are created by removing the unnecessary parts (**subtractive manufacturing**) from a block of metal, for example, by processes such as milling, grinding, drilling, and others.

  3D printing brings advantages such as the ability to create complex objects that were not possible with traditional methods and no wastage of raw material (with cascading benefits of cost reduction, alignment with sustainability objectives, and so on). On the flip side, 3D printed products aren't very robust as the various layers are glued on top of one another and stickiness will primarily depend on the adhesive properties of the raw material that is used.

- **4D printing**: 4D printing is an extension of 3D printing where, in addition to three physical dimensions, a fourth dimension of **time** is also added, which means that the structure of the 3D printed object can change with time (often under the influence of external forces or energies such as heat, electricity, light, moisture, magnetic field, and mechanical pressure). A 4D-printed object will revert to its original shape once the forces or energies are no longer acting on the product.

  Shape dynamism is achieved by using special raw materials, such as magnetic materials such as ferrofluids, Hydrogel, **Shape Memory Polymers** (**SMPs**), and **Liquid Crystalline Elastomer** (**LCE**), which have the propensity to alter their shape when some external force is applied. 4D printing is done by 3D printers but using specialized raw materials whose shape changes with the application of external forces.

Let's explore the benefits of including such technologies in IoT solutions.

Benefits of combining IoT with 3D and 4D printing

One advantage of 3D printing having direct relevance for IoT solutions is that it can be used to print a very small number of products cost-effectively. This is especially useful during the IoT solution's prototype (or proof of concept) stage as the number of devices required is small, which doesn't justify the usage of traditional manufacturing techniques.

3D printing also allows you to experiment with multiple prototype designs and select the one that is functionally and aesthetically most appropriate for the use case under consideration. In typical IoT solutions, electronic circuitry such as the **Printed Circuit Board** (**PCB**), of the field device is directly sourced from an **Original Equipment Manufacturer** (**OEM**), and 3D printing is used to quickly create enclosures for cost-effectively housing the electronic circuitry.

IoT and 3D or 4D printing can be used cyclically for successive product refinements. Consider the case of a 3D/4D printed smart glove that is sending the user's usage/comfort data to the central server. Based on the analysis of the data reported, the glove can be further refined to suit the comfort of the user. The refined design is manufactured using the 3D/4D printer, resulting in a virtuous cycle, as shown in the following figure:

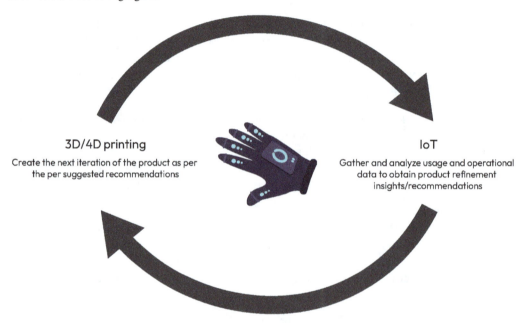

3D/4D printing
Create the next iteration of the product as per the per suggested recommendations

IoT
Gather and analyze usage and operational data to obtain product refinement insights/recommendations

Figure 12.11 – Continuous refinement of a smart product using IoT and 3D/4D printing

Next, we'll discuss some use cases in which this combination can be used.

Possible use cases

As mentioned in *Chapter 7*, 3D printing plays a crucial role in the smart manufacturing domain and especially for manufacturing hyper-personalized products. 3D printing helps us realize the vision of *batch size 1*, where each product can be manufactured in a different shape or size as per the user's preferences. 3D printing allows for *just-in-time* manufacturing and hence avoids the need to keep a large inventory of products.

IoT can be used to monitor the condition as well as output of the 3D or 4D printer. Printer monitoring would ensure that all the printing conditions such as temperature and ink availability are in the desired range. Notifications can be generated in case of product deformity while printing or can be used to notify the successful completion of the print job.

Although 4D printing is still to see widespread adoption, the technology does hold considerable potential. As an example, consider that the material that's used for the foundation of a house is 4D printed and acts as a safeguard against earthquake shocks. The 4D printed material would be able to absorb these shocks by changing shape and then reverting to its original shape.

The ability of 4D printed material to change shape or dimensions when subjected to changes in ambient temperature, heat, and more can be used to design IoT actuators that are self-powered (refer to the *Energy harvesting* section later in this chapter). Similarly, a 4D-printed actuator can be designed to perform an operation once an ambient temperature reaches a specific threshold (open a valve at a predefined temperature, for example) with the added benefit that there is no need to deploy a separate temperature sensor or a power source.

5G and 6G technology

Although **6G technology** is still in the works and **5G deployments** exist, the two technologies will be used interchangeably in this section. The main intent here is not to elaborate on the technical differences between these technologies but to understand how these two technologies would help in implementing new use cases (or enhance existing use cases). Here, it is represented as follows:

Figure 12.12 – 5G/6G technology

The reason for focusing on these two technologies compared to previous generations (2G, 3G, or 4G) is that 5G and 6G have some major improvements related to providing connectivity for IoT use cases. These improvements are summarized in the following figure:

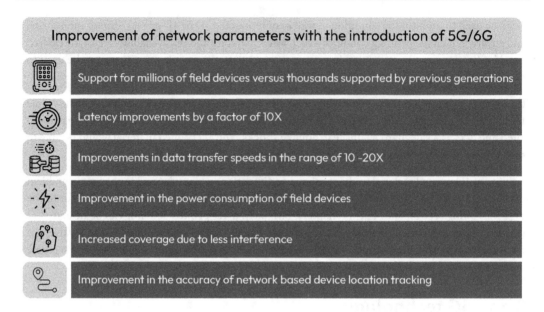

Figure 12.13 – Advantages of 5G/6G technologies over prior generations

Benefits of combining IoT with 5G and 6G technologies

One way of understanding the benefits of combining these two technologies with IoT is to consider the connectivity between the DG and the central server as a data pipe that will become bigger (more bandwidth), shorter (reduced latency – low packet transfer rate), and stronger (increased data reliability), as shown in the following figure:

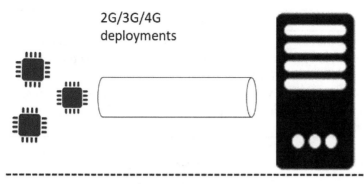

2G/3G/4G
deployments

5G/6G deployments

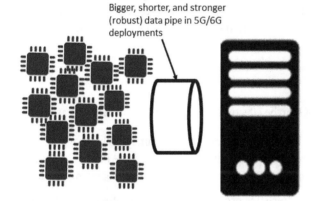

Bigger, shorter, and stronger
(robust) data pipe in 5G/6G
deployments

5G/6G benefits compared to prior
generations of cellular connectivity:
- Ultra low latency
- Bounded/predictable latencies
- Ultra reliable data delivery
- Support for dense field devices
- Better/higher coverage

Figure 12.14 – Comparison between 5G/6G and prior generations

Another important benefit of moving to 5G/6G deployments is support for dynamic network slices that allows you to create multiple logical networks within a single physical infrastructure. Each of these logical networks can be designed with specific characteristics. In other words, we can choose from a combination of characteristics that are most relevant for the use case under consideration. For example, 5G/6G allows tunable **Quality of Service (QoS)** levels for characteristics such as latency, throughput, bandwidth, data reliability, and others, as indicated in the following table:

Massive IoT	Broadband IoT	Time-sensitive IoT
Low data volume/rate	High data volume/rate	Moderate data volume/rate
High coverage	Moderate coverage	Moderate coverage
High density	Low density	Moderate density
Moderate latency	Low latency	Very low latency, but it provides bounded/deterministic latencies, irrespective of the data volume, along with ultra-high reliable delivery

Massive IoT	Broadband IoT	Time-sensitive IoT
Example use cases: smart agriculture, automated meter reading in utilities	Example use cases: autonomous vehicles, surveillance, entertainment, AR/VR applications, visual/automated inspection by drones	Example use cases: autonomous vehicles, remote surgical operations, real-time visual/audio analytics, smart manufacturing, digital twins

Table 12.1 – 5G/6G technologies can be customized to serve diverse IoT use cases

Another interesting aspect is that these characteristics can be changed on periodic time intervals or on an event basis to conserve power – surveillance cameras may share video feeds at high resolution (consuming higher bandwidth) in the night compared to the day and can be part of different logical networks in day and night.

Possible use cases

Some of the additional scenarios where 5G/6G can be effectively used in IoT deployments are as follows:

- Ability to monitor and control network status or network statistics/analytics by providing a set of **application programming interfaces** (**APIs**). Specifically, the provided APIs can be used for device management functions such as device provisioning, connectivity management, and more.

- The ability to configure network characteristics programmatically is useful for dynamically configuring the network slices, along with their characteristics (temporarily decreasing (improving) the latency of one network and offsetting that by increasing (deteriorating) the latency of some other network).

- Proactive monitoring and mitigation of security threats by promptly detecting abnormal traffic patterns.

- Dynamic switching of traffic to local DG or a central server, depending on the current network behavior.

- Some 5G variants (specifically NB-IoT) are designed for field devices that are required to consume a lot of power – the actual battery life depends on the use case and data traffic patterns. However, NB-IoT devices are typically able to run for 2 to 10 years without the need for a battery recharge or replacement.

- Before 5G solutions were available, providers relied on providing a mix of technologies to support diverse network requirements (wired connectivity for high-speed connectivity and cellular for low/moderate speed connectivity). 5G would transform these heterogeneous networks by providing a single network that would serve different needs at the same time.

Next, we'll discuss the use of drones.

Drones

For details on the features/functionalities of drones, as well as the benefit of combining them with IoT, you can refer to *Chapter 8*, in which this topic has been covered extensively. In *Figure 12.1*, it is represented using the following image:

Figure 12.15 - Drones

Possible use cases

Some of the possible ways in which IoT can be combined with **drone technology** to help develop richer/stronger use cases are as follows:

- As indicated in *Chapter 8*, drones play a crucial role in enabling smart agriculture by aiding in activities such as aerial imagery (special relevance for large farms) for monitoring crop health, planting (seeding), applying pesticides, and more to increase yield volume and quality. Drones are replacing expensive helicopters that were used for similar operations.

- IoT-enabled drones can be used to transport condition-sensitive materials such as medicines and other perishable foods/consumables while performing continuous monitoring. Drones are expected to maintain the required storage conditions of the payload and in case of any deviations/excursions, notifications are sent to the stakeholders. This scenario is similar to that of conditionally monitoring perishable goods, as detailed in *Chapter 5*. It is important to mention that condition monitoring can be done for both the drone's payload as well as the drone's operating parameters (for example, the battery's current status).

- Drones equipped with IoT capabilities can be used to inspect the assets in different domains such as utilities (inspection of substations and transmission lines over large swathes of land and hard-to-reach places). A related use case is the creation of 3D imagery using drones, where the obtained imagery can be further used for digital twin visualization.

- Drones can be used to sense connectivity challenges in an environment (no/sporadic connectivity). Field devices might be in a *sleep* state to conserve power and may not have long-range connectivity to transfer data over long distances. Drones can help in these scenarios by *waking up* the devices that are in proximity and receiving data using short-range connectivity options. Seen from this perspective, a drone can be considered a mobile DG that can collect data from remote/hard-to-reach locations with minimal to no connectivity and bring it back to the base station from where data can be synchronized with the central server.

Thus, integrating IoT with drones presents a range of benefits, paving the way for advanced automation, precision, and decision-making. Next, let's discuss how IoT can be used in social media.

Social media

Social media allows us to share thoughts, ideas, and information within virtual communities using applications such as Facebook, Twitter, WhatsApp, YouTube, Instagram, LinkedIn, and others. Social media differs from traditional sources of information as most of the content (blogs, videos, and comments) is generated by the users instead of being provided by the website owner. Also, the technology allows like-minded users to form a *network* (or a closed group) where they can comment, appreciate, or acknowledge each other's generated content. Here, it is represented as follows:

Figure 12.16 – Social media

Benefits of combining IoT and social media

Social media feeds can be used to obtain feedback regarding product launches, although the product usage metrics (which features are seldom used and which are frequently used, which time of the day the product is used, and so on) can be directly obtained from integrating required sensors into the product. The social media feeds complement these data points and provide a more holistic evaluation and future refinements of the product.

Social media can be used to determine the connections of a person who has similar preferences, profiles, nature, propensity to buy, and more and then push the related product promotions to the identified connections. Consider a scenario where a group of like-minded people are discussing the latest gadget in the market. This information can be used to send advertising or promotional messages about the related products to the group (except for the person(s) who is already using the product). Promotional messages can be further tailored as per the physical, psychological, social, locational, and economic profile of the person. This targeted marketing would be a win-win for both users as well as product providers.

Possible use cases

Social media handles can display a selected list of physiological and psychological parameters or statuses in real time that have been obtained from body sensors or wearables, providing a more holistic view of a person's condition rather than just showing just the *online/offline* status.

This dynamic physiological or psychological profile would complement the static social profile and result in much deeper engagement within social media communities/networks, giving rise to **social IoT**, where smart devices will have their own social media identities (just like humans) and they can converse with each other and with humans. As an example, one possible conversation between a connected coffee machine (as detailed in *Chapter 9*), a virtual barista, and a user via social media's messaging functionality is shown in the following figure:

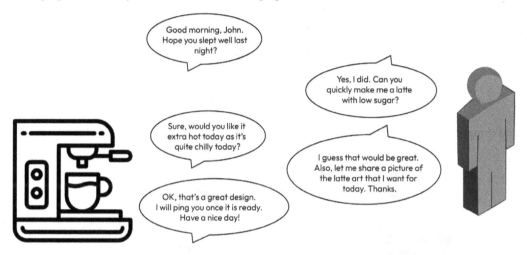

Figure 12.17 – Human-machine interaction over social media

This conversation can be between two machines, as shown in the following figure:

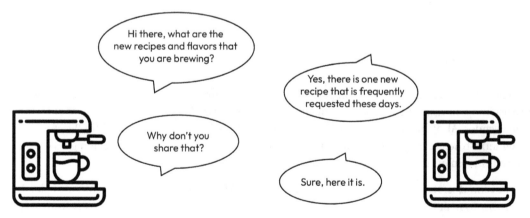

Figure 12.18 – Machine-to-machine chat over social media

As we can see, social IoT can help usher in a future where technology becomes more intuitive, interactive, and integrated into our daily lives.

Cloud computing

The cloud provides a secure, scalable, reliable, and cost-effective mechanism for providing central server functionalities, as listed in *Chapters 2* and *3*. We represent it using the following:

Figure 12.19 – Cloud computing

Cloud computing can be considered a realization of a central server, as shown in the following figure:

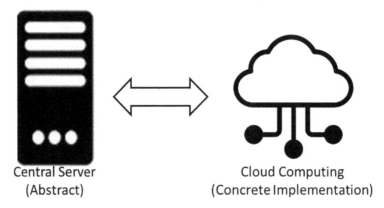

Central Server
(Abstract)

Cloud Computing
(Concrete Implementation)

Figure 12.20 – Cloud computing can be considered as a realization of a "central server"

Another benefit provided by cloud computing is that routine infrastructure activities such as hardware provisioning, infrastructure management, and software patching/upgrades are taken care of by the cloud vendor. Also, the cloud provides **Disaster Recovery/Business Continuity (DR/BC)** features as cloud infrastructure is spread globally – if the infrastructure in one geographical region goes down, it results in (almost instantaneous) automatic failover to another region.

All the central server patterns listed in the initial chapters (AI/ML integration, rule engine, file upload, enterprise system integration, device management, and digital twins) are provided as consumable services by cloud vendors. Additionally, basic infrastructure elements such as computing resources, databases, virtual machines, and more are provided as consumable services, further simplifying the process of infrastructure provisioning, which results in a considerable reduction in overall time to market.

The cloud also provides additional *plumbing* services (queuing, notifications, data buffering, and so on) that can be used to combine multiple patterns to develop end-to-end use cases. Additionally, some utility services are provided, fulfilling requirements related to caching, data transformation, device simulators, analytics, reporting, alerts and notifications, security, image or video recognition, visualization engines, **continuous integration/continuous delivery** (**CI/CD**) pipelines, developer toolchains, short- or long-term data storage, and more.

For most services, public cloud providers offer both **cloud-agnostic** as well as **cloud/vendor-native** versions. Solutions developed using cloud-agnostic features and services can be easily migrated/ported from one cloud provider to another cloud provider's infrastructure and can also be deployed on the solution provider's own data center (on-premises deployment). On the other hand, the solutions developed using cloud/vendor-native services would require considerable effort in being migrated to another provider infrastructure (or to on-premises) – often, the migration effort won't be any different from developing the solution from scratch. Like any architectural decision, selecting cloud-agnostic or cloud-native services involves tradeoffs. For example, although cloud-native services provide a lot of advantages, on the flip side, they will also bind the user to a particular cloud provider (for example, in the case of vendor lock-in). So, detailed analysis is required before deciding on whether to go for cloud-native services or cloud-agnostic services. An article written by the author of this book that provides a detailed comparison of both approaches might be a good starting point: `https://www.linkedin.com/pulse/cloud-neutral-vs-native-architects-cant-remain-fence-sitters-singh/`.

Benefits of combining IoT and cloud computing

IoT use cases generate huge amounts of data that can be effectively stored in the cloud as it provides virtually unlimited storage. One fundamental paradigm shift that cloud computing has brought compared to traditional computing is that services are charged on a usage or consumption basis and there is no requirement to make upfront payments. This is especially useful for implementing IoT use cases as most of the IoT use cases are initially developed as experimental projects or **proof of concepts** (**PoCs**) with minimal infrastructure and later scaled.

As the charging model is usage-driven, IoT solution providers can experiment with different functionalities/features, even with a small budget. In the case of increased market adoption, solutions can be scaled with a proportional increase in cost. This contrasts with the traditional approach of procuring and deploying hardware in-house/on-premises where inaccurately predicting steady-state demand results in hardware infrastructure being either underutilized or over-utilized.

Possible use cases

The seven data characteristics that are unique to IoT use cases (as mentioned in *Chapter 1*) are supported perfectly by cloud infrastructure, as shown in the following table:

IoT Data Characteristics	Supported by Cloud Capabilities
Velocity	• Scalable infrastructure to ingest high velocity/frequency data • Services designed specifically to connect with a large number of IoT devices • Buffering mechanisms in the form of queues and data/video streams to decouple ingestion from subsequent data usage • Glean real-time insights from data in transit and streaming data
Variety	• Purpose-built services to ingest and process structured, semi-structured, and unstructured data • Integration services to integrate with diverse data sources, such as enterprise systems • Support for real-time analytics as well as batch/offline analytics
Volume	• Fulfils diverse storage needs (frequently accessed data or infrequently accessed data) with corresponding cost structures • Provision to store virtually unlimited structured as well as unstructured data • Specialized databases to support IoT-specific data needs such as time series sensor data, geospatial data, and more • Services designed specifically to connect with a large number of IoT devices sending high-volume data
Variability	• AI/ML services for driving context-sensitive insights • Image recognition services for detecting "objects of interest" from audio, image, or video data • Security services provide robust authentication and encryption, eliminating the possibility of analyzing tampered/incorrect data
Veracity	• AI/ML services to automatically filter noisy/duplicate data or detect and remediate unusual traffic patterns • Certificate-based mutual authentication ensures that data is received from trusted field devices • Ensures cloud systems comply with regulatory norms (for example, GDPR, PCI-DSS, HIPAA, and NIST) to give further assurance of the data's sanctity

IoT Data Characteristics	Supported by Cloud Capabilities
Visualization	• Provides visualization services for displaying IoT data using intuitive user interfaces (for example, rolling time series data plots for sensor data) • Provides tools to develop mobile and desktop applications for consuming IoT data, insights, alarms, notifications, and more • Helps develop VR/AR applications to superimpose IoT data/insights on real-world assets/entities
Value	• The cloud provider provides sample AI/ML models related to common IoT use cases (such as predictive maintenance anomaly detection) and these can be further customized to glean insights from the accumulated data • Guidance/reference architectures specifically curated for IoT use accelerate the use case adoption and benefits realization

Table 12.2 – IoT's unique data characteristics supported by cloud services

Thus, it is evident that combining IoT with cloud computing can lead to increased scalability and security, enhanced innovation, faster **time to market** (**TTM**), as well as an overall improved user experience.

Energy harvesting

One of the main challenges in architecting IoT solutions involves optimizing the usage of power consumption in field devices. However, this challenge can be mitigated by enabling field devices to operate perpetually by leveraging perennial sources of energy (such as solar energy). Harvesting energy from natural sources not only helps in avoiding the hassle of replacing batteries and related maintenance issues (such as battery leakage) but is also more aligned with sustainability guidelines.

Energy harvesting is achieved by tapping into different energy sources such as light, vibration, motion, thermal, electromagnetic, solar, wind, and more. Some of the more novel mechanisms of harvesting energy include converting human body heat or human motion into electricity so that it can be used to power sensors attached to a human's body. Here, we represent it using the following image:

Figure 12.21 – Energy harvesting

Benefits of combining IoT and energy harvesting

Often, the combination of traditional battery and energy harvesting techniques is used to get a consistent or steady power supply. Typically, the energy produced by the energy harvesting technologies mentioned is of a small magnitude. However, this amount is sufficient for constraint devices/field devices due to their lower energy requirements.

As with all the technology choices, the right energy harvesting technique will depend on the use case or application needs (frequency of data collation, transmission by field sensors, transmission range, the device's active and sleep cycles, the mobile or stationary nature of field device, and so on), cost considerations, and ambient or environmental conditions.

Possible use cases

Energy harvesting can be used to deploy IoT field devices in places where normal power sources (such as electricity) are not available and battery replacement is also not an option. Energy harvesting can be used in remote locations to provide continuous power to field devices (for example, conditionally monitoring crops in remote agricultural land).

Quantum computing

Quantum computing provides a drastic improvement in computing power compared to the current generation of hardware. Quantum computing technology operates by encoding the data in the form of qubits (quantum bits) and can store non-binary data (states other than 0s and 1s). Traditional computing infrastructure could store and process binary data only. Information stored in qubits allows quantum computers to solve extremely complex mathematical/statistical problems in a very short time. For example, in 2019, Google announced that its quantum computer was able to solve a complex problem in 200 seconds, which would have taken a traditional computer infrastructure about 10,000 years to solve.

We represent it using the following figure:

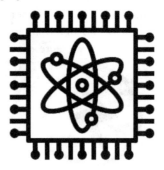

Figure 12.22 – Quantum computing

Like any other technology, quantum computing can be exploited for malicious intents and the one possible threat is that it can be used to break currently used passwords and cryptographic keys owing to its immense computing power (it would take a key break algorithm hundreds of years to break a decent length cryptographic key on a traditional computer, but this can be broken in minutes if the same algorithm is executed on a quantum computer).

At the time of writing, quantum computers are very expensive and used for very specific purposes only. Quantum computers also need to be housed in extremely cold temperatures, which increases the cost and complexity of their deployment. However, quantum services can be rented from cloud providers by following their *pay-as-go* billing model (refer to the *Cloud computing* section of this chapter), at which point they can be used to execute computationally complex applications and algorithms.

Benefits of combining IoT and quantum computing

Quantum computing doesn't have much of a role to play in the case of field devices (complex computing is not required on the field/constraint devices and it is almost impossible to replicate the specialized environment that is needed for quantum computers to operate). However, quantum computing can be used to implement very special use cases at the central server that require tremendous computing power (digital twin simulations, complex analytics, process optimizations, complicated AI/ML models, and more).

Possible use cases

With time, the amount of data that can be generated by field devices will increase exponentially (as more and more diverse devices get connected) that can be easily processed/analyzed by the computing power provided by computers that use quantum technology.

Web 3.0

Web 3.0 is the third stage of the evolution of the World Wide Web or the internet and has been designed to overcome the limitations of prior stages – that is, Web 1.0 and Web 2.0. In *Figure 12.1*, it is represented as follows:

Figure 12.23 – Web 3.0

Before we understand Web 3.0 and how it can be combined with IoT, let's first understand the web's evolution from Web 1.0 to Web 3.0, as depicted in the following figure:

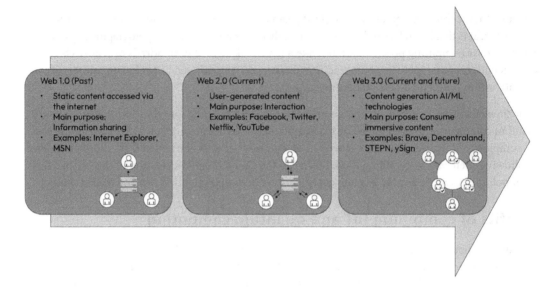

Figure 12.24 – Evolution of the web – from Web 1.0 to Web 3.0

Let's look at this evolution in greater detail:

Parameter	Web 1.0	Web 2.0	Web 3.0
Content source	Static content accessed via the internet	User-generated content	Content generation, moderation, filtration (identification of fake data), and curation powered by technologies such as blockchain, **non-fungible tokens** (**NFTs**), cryptocurrency, and more. Additional attributes include permissionless (no user can be banned or restricted), trustless (two or more users can trust each other without relying on a third party), and support for data monetization.
Type of access	User access allowed: read-only	User access allowed: read and write	User access allowed: read, write, and own.

Parameter	Web 1.0	Web 2.0	Web 3.0
Content type	Simple web pages	Dynamic web pages, social media	Dynamic content consumed using immersive technologies and real-time data ingestion from IoT sensors.
Key purpose	Information sharing	Interaction	Consume immersive content.
Type of client consuming content	Desktop browsers	Mobile phones, touch screen HMIs	Immersive technologies such as AR/VR, the metaverse, and others.
Type of storage used	Dedicated infrastructure/ data centers with limited scalability options	Distributed data centers (cloud infrastructure) with full scalability	Distributed data centers (cloud infrastructure) with full scalability, along with edge locations.
Content moderation	Performed manually	Performed manually	Performed using AI/ML tools.
Data ownership	Content provider	Content/application provider	(End) user.
Typical examples	Internet Explorer, MSN	Facebook, Twitter, Netflix, YouTube	Brave (browser), Decentraland (gaming), STEPN (health and wellbeing), ySign (chat messenger).

Table 12.3 – Key differences between Web 1.0, 2.0, and 3.0

As detailed in the preceding table, Web 3.0 is the latest stage of the web/internet's evolution and is differentiated from the earlier stages in the following manner:

- **Decentralized/distributed**: Content isn't owned and/or controlled by a few organizations but ownership is with the content creators and users. In Web 1.0 and Web 2.0, a user's account was owned by the entities providing the online platform (for example, email, online gaming, and so on) and they could suspend or even terminate the account (or restrict specific content) as per their needs and requirements (for example, to comply with regulatory or legal directives). Also, the data in Web 1.0 and Web 2.0 was managed in a central location, which posed a security risk as bad actors could steal the data of all the users by exploiting a single vulnerability.

Web 3.0, however, allows you to purchase part of the platform (virtual real estate) using **non-fungible tokens (NFTs)**. Once purchased, the user is free to use, sell, or rent in the open market as per their wishes. In other words, once purchased, the user would own part of the platform, not very different from owning shares in a publicly listed company. Powered by blockchain technology, Web 3.0 provides a level playing field for content creators where they are not bound by enforced censorship and arbitrary rules.

- **Cryptocurrency as a mode of payment**: Payment for any online purchases on Web 3.0 is done using cryptocurrency in contrast to traditional payment methods such as credit cards, a bank's online payment gateway, and so on.

- **Limited or no trust boundaries**: Trust is not guaranteed by trusted third parties but is enforced indirectly by other economic incentives or penalties.

- **Permissionless**: Anyone can participate in content generation and consumption without the need to enter legal contracts.

- **Democratized**: The feature enhancements that have been made to Web 3.0 apps are decided by all the participants by a majority vote. This contrasts with Web 1.0 and Web 2.0, where application providers determined the features to be implemented based on their understanding and preferences.

Web 3.0 can also be considered a combination of technologies, where IoT is responsible for gathering real-world data, as shown in the following figure:

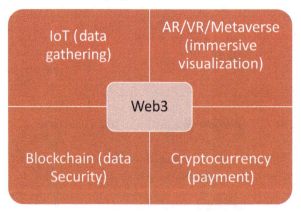

Figure 12.25 – Web 3.0 leveraging additional technologies

Benefits of combining IoT and Web 3.0

Users can monetize the data and insights generated by IoT applications securely as the financial transactions are secured by NFTs and cryptocurrencies. Like other technologies listed in this chapter, IoT and Web 3.0 complement each other's functionality, where IoT is responsible for gathering data from the real world and deriving meaningful insights, whereas Web 3.0 helps democratize the accumulated data and insights and enables users to consume data and insights immersively, along with the required safeguards and controls.

Normally, IoT deployments are architected in a **centralized** manner, whereby multiple field devices push data to a central server where it gets aggregated and required insights are generated. Accordingly, the decentralized nature of Web 3.0 can be used to experiment with additional architectures where field devices can communicate with each other by using smart contracts (enable by blockchain) without the need for a central server (provided that the field devices are capable enough to run processing, such as blockchain operations).

Combining both technologies would also enable the owners of IoT data to monetize their data in the marketplace without involving any intermediary (considering the use case of land consolidation from *Chapter 8*, agriculture landowners would be able to trade data and insights without any third-party involvement). Also, the payments for such transactions can be made securely using cryptocurrencies.

Another benefit that Web 3.0 can bring is that it can provide **immersive visualization**, using which users can understand the data, insights, and the related context intuitively (refer to the *Immersive technologies* section in this chapter). Also, the decentralized nature of Web 3.0 can be used to share data and insights between field devices rather than aggregating all the data to a central server. Blockchain (which is one of the foundational technologies of Web 3.0) can provide authentication between these field devices, as mentioned earlier.

Web 3.0 helps democratize the consumption of data and related insights by developing immersive applications over the internet, whereas IoT helps transfer that data over the internet from field devices to a common aggregation point.

Possible use cases

Combining Web 3.0 with IoT would allow ownership as well as monetization of data and insights. Consider the case of smart homes, where individual homeowners have smart energy meters, which they can use to track the consumption of different appliances or equipment in their homes. Web 3.0 allows the ownership of not only hardware (that is, smart meters, home automation devices, and so on) but also the data generated by these devices (end-to-end ownership model).

Homeowners can then sell their data to interested parties (possibly to the highest bidder). Additionally, owners can visualize the energy consumption-related data and insights for their homes (along with recommendations for saving energy) immersively (data overlaid on the actual equipment or appliance).

Homeowners may also opt to give back the energy generated using harvesting technologies such as solar to the grid and receive payments securely. This use case can be further extended whereby a large (and expensive) asset (such as a windmill) is jointly owned by multiple homeowners. Here, Web 3.0 allows homeowners to securely manage ownership records, as well as fairly distribute generated income.

Edge computing

Edge computing refers to the deployment architecture where data processing is done near the point where data is being generated. In the context of IoT, the processing done at DG (instead of sending the data to the central server) is a good example of edge computing in action. This is primarily required in scenarios where quick data analysis and action are required (the best example would be autonomous driving, where even a split-second delay could have serious implications). Edge computing also aids in ensuring privacy and security since data is stored and analyzed locally – a good example of *security by obscurity*. We represent it using the following:

Figure 12.26 – Edge computing

An important point to note here is that having edge computing doesn't imply that no data is flowing into the central server. Extending on the autonomous vehicle example, although most of the processing/decision-making (analyzing the feed from multiple onboard sensors to determine whether to accelerate or to slow down) would happen at the edge, the consolidated/summarized data about these decisions can be sent to the central server for aggregation, data storage, and model refinement.

Some additional applications where edge computing plays an important role are as follows:

- Video analytics
- Automated quality control on an assembly line
- Driver behavior monitoring and control
- Health monitoring and dosage control systems

> **Important note**
> You are encouraged to review the DG pattern (elaborated in *Chapter 2*) to better understand how DG implements edge computing and scenarios where edge computing is relevant. Edge analytics was also covered in detail in *Chapter 10*.

Benefits of combining IoT and edge computing

Edge computing is not separate from IoT technology (DG being responsible for edge computing is a crucial element of the IoT reference architecture) and there are numerous advantages of processing the data at the edge rather than at the central server. Some of those scenarios are listed here:

- Edge computing is expected to generate data insights with minimal latency by avoiding a round trip to the central server. Although internet speeds are increasing with time, most long-range data transfers happen over fiber networks, where signal transfer is limited by the speed of light.

- In some cases, local processing also helps conserve the precious battery power of the constraint devices as there is no need to establish and maintain connectivity with the central server. Also, often, pre-processing or formatting data or protocol translations is required to make the data compatible with central server requirements. This extra processing also consumes decent battery power. In some remote locations, connectivity to the central server may be intermittent or even nonexistent. This also necessitates processing the data at the edge.

- Another reason for performing edge computing is to conserve channel bandwidth (as we have seen in past chapters, normally, video analytics is done locally (that is, at DG) and the processed results (local video analytics can determine if there is any action or movement in the video feed in real time) are sent to the central server for aggregation or to trigger additional business rules).

- In some scenarios, not sending the data to the central server and relying on edge computing can be due to governmental regulations that restrict the movement of data outside the physical boundaries of the specific geography or country. Some organizations prefer local processing/edge computing to mitigate their concerns about data security. Normally, the central server (for example, a cloud server) is hosted on the vendor's premises, which can result in data security breaches. Also, as seen in *Chapter 11*, data transmission to the central server and back results in an additional set of security risks. Additionally, as data is aggregated in the central server from multiple DGs, the potential impact of a security breach at the central server is much higher than at the individual DG level. These concerns, coupled with the desire to protect sensitive/private customer data, often force organizations to rely on edge computing.

The benefits of edge computing are summarized in the following figure:

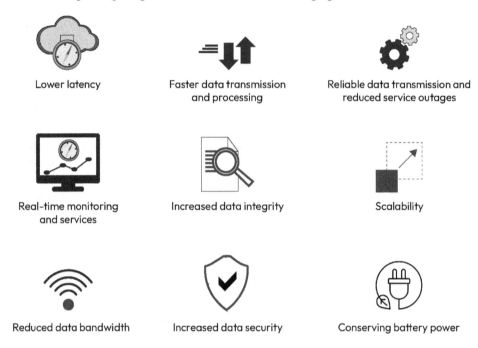

Figure 12.27 – Benefits of edge computing

The distinction between local and remote processing is subjective and there can be intermediate hops (Intermediate server) for data processing/aggregation before it finally reaches the ultimate aggregation point (central server), as shown in the following figure:

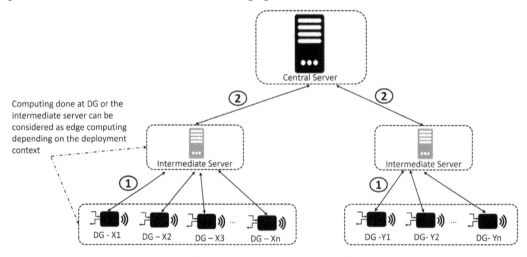

Figure 12.28 – The interpretation of edge computing varies depending on the deployment context

As stated earlier, edge computing refers to the scenario where data processing is done near the point of data generation. Considering this, processing done at DG should be considered edge computing. However, if the physical distance between *DG* and the *intermediate server* (marked as **1**) is much smaller compared to the distance between the *intermediate server* and the *central server* (marked as **2**), even the computing done at the intermediate server can be considered edge computing.

> **Important note**
> Processing done at the intermediate server is also referred to as **fog computing**.

In addition to these benefits of edge computing, another benefit that deserves special mention is that field devices (smart sensors and DGs) can transfer the local processing needs to the edge or intermediate server, thereby reducing the hardware complexity, cost, and power requirements of a large number of field devices.

> **Important note**
> Due to the physical proximity of 5G base stations to the field devices, these base stations are also often used as intermediate/edge servers.

Possible use cases

IoT solutions and applications where processing is done near the point where data is generated are all examples of edge computing. This includes autonomous vehicles, video surveillance, in-hospital patient monitoring, and faster content delivery (for example, video streaming applications). In addition, common IoT features that require local processing such as image identification (for example, facial recognition), gesture or voice recognition, motion detection, and more can be effectively performed at the edge or on an intermediate server.

For in-home automation use cases, edge processing can be used to detect the presence of an intruder and send an alarm to the homeowner.

The processing that's required to provide sensor fusion (inputs from multiple sensors are combined to generate firmer sensing information) is also generally performed at the edge or on an intermediate server.

This was the last example demonstrating IoT technology being used in conjunction with complementary technologies to reap additional benefits. As mentioned earlier, the idea here is not to provide an exhaustive list of technologies but to emphasize the need to continuously scan the market for emerging technologies and evaluate their relevance in the context of IoT. Before closing this chapter, let's quickly recap what we learned.

Summary

This chapter provided a representative list of technologies that can either *complement* (provide additional features or functionalities, such as 3D/4D printing) or *supplement* (provide nonfunctional capabilities, such as performance enhancement using quantum computing) IoT technology. This chapter should help you, the solution designer, to evaluate the problem at hand in a more holistic manner and determine what parts of the problem IoT can solve and where there is a need to include additional technology.

Some of the technologies described in this chapter are relatively new, whereas others are evolved/mature technologies. One important takeaway from this chapter is that when solving a complex problem, it is better to focus on the capabilities needed to solve the problem and then map the identified capability to the technology (which can be either a new or matured technology) – often, the latest technology might not be the ideal fit for the problem at hand. You are encouraged to look out for and spot other technologies that are not listed in this chapter but can be effectively combined with IoT (for example, robotics) to provide more comprehensive solutions.

Although this chapter showed how combining a specific technology helps augment IoT capabilities, it does bring one important additional point – the technologies listed in this chapter are building blocks (akin to Lego blocks) that can be mixed and matched with IoT in any manner to create richer solutions that help solve complex problems. As an example, IoT, 5G and 6G, edge computing, and AR/VR can be combined to develop a more holistic solution where requirements such as faster local processing, low latency complex computations on the central server, and generated insights need to be consumed immersively using an AR/VR headset in a single solution.

The final chapter of this book focuses on offering practical guidance for implementing IoT solutions, while also sharing some insights gained from hands-on experience in this field.

13
Epilogue

As I begin to write the final chapter of the book, I am experiencing mixed feelings – satisfaction about this book soon hitting the stands and anxiety about whether it will reach its intended audience. As we wrap up this book, the most obvious way forward in this chapter was to summarize the key points of what has been covered. However, as this might have limited benefit for you, I ultimately decided to jot down the key takeaways I have gathered while working on IoT solutions and projects for nearly a decade. Providing practical tips drawn from actual project experiences will also help to balance the relative abstractness of a few of the prior chapters. Accordingly, this chapter lists a few nuggets of knowledge that might help you to avoid common pitfalls or mistakes in IoT implementation.

Most of the software engineering principles and best practices that ensure the successful completion of a project are applicable to IoT projects as well. However, IoT solutions are unique as (by definition) they span both physical and virtual worlds, resulting in additional (technical as well as project management) nuances that should be carefully considered to deliver a successful IoT project. As a result, the chapter also lists a few project management considerations.

As we all know, a project/solution can't be labeled as successful if it meets the functional requirements while ignoring performance or other similar **non-functional requirements** (NFRs). Again, this is not specific to IoT solutions but the impact of ignoring NFRs is relatively high for IoT implementation – as an example, consider the simple case of the delay of a water sprinkler when a fire sensor has already detected fire. Accordingly, we will discuss IoT-specific NFRs in the *NFR considerations* section.

Finally, no book can be complete without listing the possible connectivity options/protocols (along with their associated trade-offs) needed to connect field devices with the central server. So, the last section lists the possible connectivity options/protocols along with recommendations for which option is suitable under what circumstances.

Let's start by exploring some key considerations for effective project management.

Project implementation considerations

The points listed in this section may appear to be more relevant to a project/program manager; however, they would be useful for architects as well, since architects are expected to remain part of the implementation team even after the completion of the architecture. The expectation stems from the fact that the architect bears the responsibility of ensuring that the implementation is always aligned with the approved architecture. So, the following project management/implementation considerations for IoT projects would benefit people belonging to technical and project management roles:

- **Component procurement considerations**: Field devices (sensors, gateways, and actuators) and components providing connectivity (such as SIM cards) form a key part of any IoT solution, and the timely procurement of these components is crucial. However, the procurement and deployment of these third-party components pose a challenge, as multiple parties (such as system integrators, device vendors, device compliance vendors, and security experts) are involved. Moreover, the solution provider/system integrator has limited control over these component vendors. In fact, each vendor or third party has its own lead times or **Service-Level Agreements (SLAs)**, which make the delivery of the overall IoT project quite unwieldy. In other words, hardware components (a physical entity) must go through a complex supply chain before they reach the hands of the solution provider or system integrator.

  These are a few examples of complex interdependencies that make the tracking and management of project schedules quite difficult and cumbersome. Project or program managers can mitigate this risk effectively by clearly identifying dependencies upfront and then monitoring those dependencies at regular intervals.

  It is also equally important that a detailed analysis of the vendor's ability to deliver parts on time is done in a timely manner (preferably at the start of the project) to avoid unpleasant surprises later. A detailed review of short-listed vendor contracts (especially terms and conditions related to liability clauses, compliance/validation mechanisms, SLAs, and so on) should be done alongside the organization's legal teams. As compliance/regulatory norms vary from one geographic region to another, the vendor should be willing to provide the compliance certifications for all the regions where the solution is expected to be deployed.

  Quality- and performance-related risks can be further mitigated by requesting sample hardware pieces from the vendor and by testing them in-house before placing bulk orders. Another point that deserves scrutiny is the ability of the vendor to fulfill bulk orders (volumes in the range of tens of thousands to millions) and that too within required timelines.

  In my experience, I have seen vendors that can easily deliver a small number of devices/components but struggle to fulfill bulk orders. In fact, in one of our *asset tracking* projects, we realized very late in the project life cycle that the vendor wasn't equipped to provide the number of RFID tags that were needed to tag all the parts that were required to be tracked as part of the contract, although the vendor consistently made claims to the contrary. This forced the team to look for an alternate vendor, which wasted crucial time. One effective strategy to prevent this type of situation is to onboard multiple vendors or partners to supply crucial components, thus avoiding single-vendor dependency.

- **Verification/validation considerations**: There are some unique points to consider while performing validation or verification activities for IoT projects:

- Some verification/validation steps are needed only for hardware components, and as such, the team may not consider those while finalizing the verification/validation scope. As an example, hardware components (especially sensors) show a gradual drift in their operation and require constant validation/recalibration.

- It is almost impossible to replicate real-world scenarios in the lab (for example, battery characteristics, connectivity issues such as limited or erratic connectivity, and load scenarios such as 10,000 sensors sending data to a central server) and the verification engineers need to rely on field testing, which is only possible near the *go-live* date.

These examples indicate that a structured testing approach tailored to IoT-specific use cases needs to be considered. Some of the key elements of such a structured approach (different testing/verification types that are relevant for IoT use cases) are shown in the following figure:

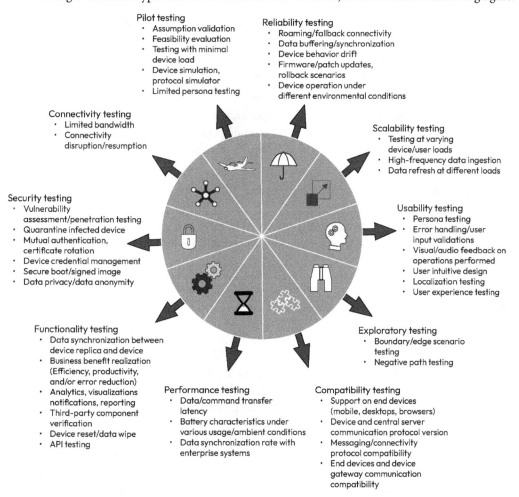

Figure 13.1 – Types of recommended IoT testing techniques

- **Simulator-specific considerations**: Often, the requirements can't be met with *off-the-shelf* hardware and the hardware components need to be custom developed, and that too needs to be in parallel to the other solution components (for example, DG communication with the central server needs to be validated even when the DG is being developed or validated). Software simulators can help in filling this gap; however, this results in additional complexities of *build versus buy* decisions – whether to build simulators in-house (more control and protection of intellectual property) or procure them from the market (less control over delivery timelines along with risk to proprietary information) – or the simulator's functional scope (whether the simulator is intended to only replicate the device behavior or it is required to simulate internal logic as well). Also, spending too much time on developing or validating the simulator may not be advisable as it would be a stop-gap arrangement until the actual hardware was available unless it is to be used during the validation/verification phase to perform load tests.

As can be seen, the usage of simulators brings forth additional complexities/dependencies that need to be carefully analyzed. The capability and accuracy/sophistication of simulators need to be given due importance during the requirements elicitation phase. As an example, for each simulator, the following needs to be mandatorily considered:

- **Type of functionality to be simulated**: Whether the simulator is required to simulate only the communication with other components or it will be required to simulate internal characteristics (for example, power/battery consumption, consistent/intermittent connectivity, and/or latency scenarios). Similarly, whether the simulator would be testing only the *happy* scenarios or the erroneous scenarios as well.

- **Level of intelligence inbuilt into simulator**: Will the simulator's internal logic be based on a simple set of rules (if the input is X, return Y) or will it involve complex intelligence (considering prior temporal context, calibration drifts, and so on)? In the latter case, additional AI/ML capabilities need to be considered (as detailed in the AI/ML integration pattern listed in *Chapter 3* and *Chapter 10*).

The section listed some IoT-specific project management considerations and guidelines. In the next section, I will share some lessons learned from my past IoT projects.

Lessons learned from IoT projects

There are a few lessons that I've learned from my experience of working in the IoT domain. Keeping in mind the list of miscellaneous points mentioned in this section may help you avoid unnecessary hassles/issues. The list of the key lessons (in no particular order) is as follows:

- **Scoping a project with unclear objectives or an unclear end state**: There are cases where the scope of the IoT solution is clearly defined and there is clarity on the overall business objective. However, there are other scenarios where practitioners struggle to find the right-fit or optimum solution for the stated business problem. A limited-scope **proof of concept** (**PoC**) or pilot is typically executed to reduce the ambiguity and number of unknowns. As the budget for the

PoC/pilot is limited, there is a need to optimize scope/cost but still obtain crucial inputs that can help to move in the right direction.

In addition to technical/architectural skills, all this requires the ability to think creatively and bring to bear an innovation mindset to generate the right set of implementable ideas. One technique that is really useful in this regard is the **Theory of Inventive Problem Solving** (TRIZ) as it can help improve the odds of success in a systematic manner rather than relying on the *trial-and-error* approach (for more information, refer to `https://www.mindtools.com/amtcc5f/triz` and `https://www.ee.iitb.ac.in/~apte/CV_PRA_TRIZ_INTRO.htm`). In general, TRIZ is useful for finding a creative solution to any engineering problem; however, it's more relevant in the IoT context as the problem (and solution) space transcends both physical and virtual realms. Going into more detail about the technique is beyond the scope of this chapter; however, you are highly encouraged to explore this useful technique whenever there is a need to explore innovative solutions for the stated problem.

- **Understanding the explicit and implicit requirements**: As IoT is an emerging technology, not all users may be aware of the capabilities and benefits that this technology can provide. As a result, it becomes important for the solution providers to fully understand the stated and unstated (implicit) needs. It is possible to gauge these requirements from the provided documentation or by doing customer interviews. However, a better approach to understanding the customer's operating context and their problem domain is by making physical site visits to their premises. These visits provide the opportunity to understand the constraints under which the solution would be expected to operate – for example, constraints such as limited connectivity, reliance on non-conventional power sources (solar), and so on. These onsite visits can be made more efficient when the solution provider visits the site fully equipped with a set of questions and/or assumptions. These visits can also be used to test the actual operation conditions by asking questions such as, "Is the SIM provided by the connectivity provider able to connect with the central server from different geographical regions?"

- **Early and continuous feedback from stakeholders**: Often, it is difficult to clearly articulate the final or desired outcome due to factors such as hardware/software integration, connectivity challenges, power availability for field devices, and so on. Sometimes, there are even doubts about the viability or feasibility of the solution. Both these challenges can be effectively handled by executing the IoT projects using an **Agile methodology** (in contrast to the traditional Waterfall model) as it is better suited to handling the risks, ambiguity, and uncertainties mentioned. Agile methodologies such as Scrum, Kanban, and the **Dynamic System Development Method** (DSDM) also force continuous interaction between the solution provider and consumers, which gives the consumers an opportunity to provide early feedback while the solution is still being developed, resulting in fewer surprises later. Continuous feedback also ensures that the overall solution is not overly complicated or over-engineered while still meeting all the requirements. Although early and continuous feedback (one of the key features of Agile methodologies) is a crucial factor that helps in managing expectations in any project, its relevance is even more important for IoT projects because of its broader and varied scope, as it includes hardware and software elements. In fact, feedback needs to be taken regarding software requirements – including the look and feel of the **user interface** (**UI**) and dashboard elements, the type of data to be collated and analyzed, and business process flows – and consumers need to be aligned on hardware requirements, such as dimensions, ruggedness, and aesthetics of physical components (for example, field devices or gateways). Hardware components are much less malleable (adaptable) to changes (especially those requested toward the end of the project life cycle) compared to software components. Here are some examples from my personal experience:

- I recollect an incident where during the **User Acceptance Test** (**UAT**) phase, a customer raised the concern that affixing the IoT tracking tags to their assets would impact the overall aesthetics and they wanted the tags to be replaced with more compact tags. This was the case even when the customer was shown the pictures and the dimensional specifications of the tags were also shared. Obviously, it was difficult to replace tags toward the end of the project life cycle. In hindsight, it would have been better to share actual (sample) tags along with specifications/documents, as people have different perceptions when having a tag/device in their hands (touch and feel factor) vis-à-vis just looking at pictures/specifications.

- In another project, a customer refused to accept the project delivery as the RFID tags that were used to track their IT assets were not aesthetically appealing.

- In another case, a customer wasn't satisfied with the DG's overall dimensions, although the gateway was satisfying all the other functional/non-functional requirements.

- **Early identification and tracking of business objectives**: IoT projects are conceptualized and implemented to achieve certain business objectives or to solve a particular business problem. Some of the common business objectives that are relevant to IoT projects are listed in the following figure:

Figure 13.2 – Representative list of IoT solution's business objectives

It is important to document such objectives as *success criteria* against which the project's performance would be evaluated. One point you will have realized is that these business objectives are different from the solution's features and functionalities that are recorded during the requirement/scope finalization phase. It is possible that a solution implements all the requirements (features/functionalities) and is still not able to meet any of the business objectives. Hence, it is important to keep an eye on the overall objectives during the entire project life cycle in the context of requirement analysis, architecture, design, implementation, verification/validation, and deployment.

- **Agreement on the parameters to define solution value**: One important framework that can be used to elucidate value to the customer is **Bain's framework** (https://media.bain.com/elements-of-value/#, https://hbr.org/2016/09/the-elements-of-value). This framework helps to articulate value from diverse perspectives such as emotional, social, and other similar needs. It is recommended to use this framework to list all the possible benefits that the customer is expected to gain once the product is released and then cross-validate them with the customer. Another similar and equally powerful framework to determine/articulate the value across a complete supply chain is **Porter's Value Chain analysis** (`https://www.smartsheet.com/value-chain-model`).

- **Following a change management process to address cultural/personal sensitivities**: The deployment of the IoT solution will bring about changes in existing ways of working and hence like any other change management initiative, the IoT solution implementation will involve human/cultural sensitivities that need to be suitably addressed or managed. Although delving deeper into the point would be outside the scope of the chapter, it is important to mention that the whole IoT initiative may get derailed if adequate focus is not given to people directly impacted by the change. Resistance to moving away from current ways of working is one of the main reasons for people not accepting or adapting to them. This can be circumvented by taking measures such as clear articulation of need as well as benefits of change, involving people impacted by the change in the change management process, and avoiding the trap of change for the sake of change.

- **Data as a decision enabler**: It is important to ensure that data collected by using IoT solutions is being used for decision-making and is not collated just for compliance purposes. Also, IoT solutions provide a unique opportunity where solutions can be suitably instrumented, resulting in the capture and analysis of product usage data. This allows for determining how (or whether) the product is being used in the field, and the collated product insights can be used for future product refinements.

- **Validation of architecture and architectural decisions at regular intervals**: Key architectural decisions should be made as early as possible so that there are minimum changes to the downstream project activities. This may sound obvious, but most IoT projects start as PoCs or **proofs of value** (**PoVs**), which are limited in both scope and scale, and the focus is to ensure the use case's feasibility.

Ideally, the architecture for the PoC/PoV should be revalidated before implementing the full scale-out (production-ready) solution. In fact, some consummate architects suggest discarding all the deliverables that are created during PoC/PoV stages (as they are created with minimal focus on quality as well as solution longevity) and developing the deliverables (architecture, source code, and so on) from scratch once the project is approved for full-scale implementation.

As an example, in one of our projects, the team started with a monolith architecture for implementing central server logic (at the PoC stage) as there was much less confidence in the project outcome and the team wanted to keep the architecture simple for the initial phase. However, once the PoC's results turned out to be positive, the team had to change the architecture to a microservices-based architecture, resulting in considerable refactoring effort.

- **Alignment with an API first strategy**: This strategy allows crucial functionality to be exposed via a set of APIs. It enables users to consume IoT data/insights from a diverse set of clients (mobile, wearables, desktop, and so on). In the IoT context, following the *API-first* approach has an additional benefit as it allows for easier integration with enterprise systems (as mentioned in the *Enterprise system integration* section in *Chapter 3*).

- **Data lake architecture for ingestion/analysis of diverse data types**: Most of the use cases discussed in this book (*Chapters 4* to *8*) have provided details about how architectural patterns can be used to realize mentioned use cases, and most of the use cases target one specific requirement/problem statement. However, for the implementation of complex use cases (one example can be a land trading platform, as detailed in *Chapter 8*), where diverse data needs to be aggregated and analyzed (including unstructured data such as video imagery, semi-structured data such as data files from enterprise systems, and structured data such as sensor data), a data-lake-based, platform-centric approach would be preferable. Although covering data lakes in detail is beyond the scope of this chapter, the point here is that for some of the complex use cases where data is expected, the data lake architecture is one of the good options to consider.

- **Choosing between cloud-native and cloud-agnostic IoT services**: One of the key architectural decisions that needs to be taken upfront when we intend to leverage the public cloud for central server implementation is regarding whether architecture should leverage vendor-specific services (often called *cloud-native* services) or rely on open source components/services (known as *cloud-agnostic* services). This decision is important as it has a long-term bearing. A few pointers that can help in reaching optimum decisions are as follows:

  - Is the solution expected to be portable – that is, migrated from one cloud provider to another in the short or medium term?

  - How do the IoT-related services (such as services related to data ingestion, stream processing, device management, analytics, and so on) of the selected vendor compare with other vendors? I worked on a project where one service provider used all services from one vendor except the analytics service, where he was relying on another vendor – a situation that resulted in a complex multi-cloud deployment.

  - How do the development or deployment time costs of IoT and related services compare between different vendors? Typically, the price of the cloud-native services is kept lower to prevent users from moving to other cloud providers' ecosystems.

  - The complexity of the architecture when deployed using open source services compared to deployed using cloud-native services. The architectural complexity is less profound in cloud-native solutions compared to cloud-agnostic solutions. Additionally, steady-state development and/or enhancements can be easily done in cloud-native environments as most of the cloud provider tools are well integrated.

- **Solution deployment/installation considerations**: One often overlooked but crucial phase requiring detailed planning (and close monitoring) is the *field deployment* phase, where field devices are deployed/installed in the field. To avoid unpleasant surprises (also toward the project completion when time is really of the essence), it is recommended that the customer is adequately informed regarding various deployment-related requirements (such as power and/or connectivity requirements) so that they can make relevant arrangements in advance. Here are some examples:

  - If the solution relies on **Power over Ethernet** (**PoE**), the customer would stand to benefit from or prefer prior intimation so that they can start the required cabling/masonry work well in advance.

  - Similarly, sensor (or any other field device) installation at a crowded/public place can be efficiently and effectively done during off-peak hours (such as during the night). As this might involve multi-party support (device vendor, solution provider, customer, local authority, and so on), close coordination, as well as planning, would be needed among all the involved parties.

The section listed quite a few guidelines and recommendations that can enhance the chances of successful IoT deployments and help us to reap the desired benefits/business objectives. Like any other solution development endeavor, implementing functional requirements is just one part of the story and doesn't necessarily guarantee satisfied stakeholders, and it is equally important to give sufficient attention to NFRs as well, which we will cover in the next section.

NFR considerations

Most of the previous chapters focused on implementing **functional requirements**; however, NFRs (especially the ones that are related to the solution's efficiency or performance) are equally important. Not giving required importance or priority to non-functional aspects (such as battery life, data transfer latency, and compatibility with different connectivity providers) can adversely impact the customer's experience, jeopardizing the realization of business objectives.

All the components of the IoT system should be optimized individually to provide overall (end-to-end) efficiency; however, performance expectations are generally more stringent for field devices than for central servers (for example, frequent replacement of batteries in remote or large geographical areas is difficult from a cost as well as an operational standpoint). Accordingly, some of the key NFRs that should be considered while conceptualizing/developing IoT solutions are listed in the following figure:

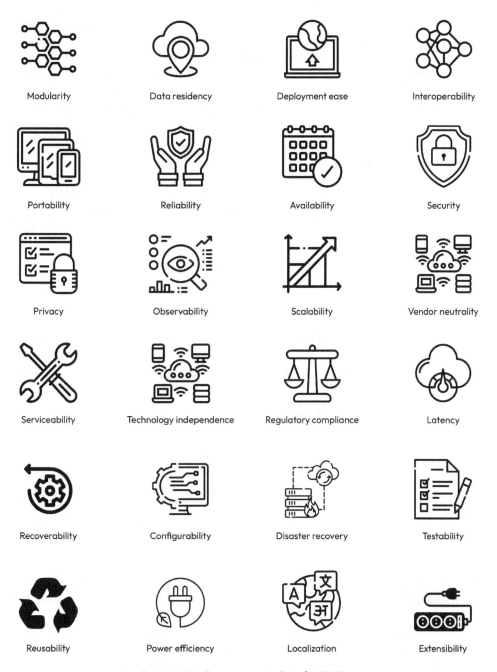

Figure 13.3 – Representative list of IoT NFRs

To realize each of the NFRs listed in the previous figure, due consideration should be given during the architecture phase. As an example, if bandwidth optimization is one of the expected NFRs, it would impact architectural decisions as well as technology choices. For example, using less chatty protocols such as **Constrained Application Protocol (CoAP)** over **User Datagram Protocol (UDP)** is much more efficient than using HTTP over **Transmission Control Protocol (TCP)** and helps to optimize bandwidth consumption. Providing details about every NFR listed in the previous figure would be beyond the scope of this chapter; however, this section does list some tips/techniques that can be used for optimizing two representative NFRs – **battery optimization** or **power optimization** and **cost optimization** of the solution development.

Battery or power optimization

The amount of battery or power a field device can consume (known as the **power budget**) has a direct bearing on a host of other architectural decisions/choices, including the following:

- Type of connectivity protocol

- Type (raw or processed data) and frequency of data transfer

- Payload size

- Processor selection

- Type of enclosure (battery consumption is impacted by ambient conditions such as temperature and humidity – higher ambient temperatures reduce the battery discharge rate and can power the device for a longer duration, and the presence of electromagnetic interferences has the opposite impact on battery life)

- Physical dimensions of the field device (in general, batteries with bigger sizes can support longer operation)

- Level of security hardening (type/rigor of security algorithms supported)

Increased public awareness and focus of regulatory bodies on aspects such as sustainability and reducing e-waste in the recent past ensures that power optimization is no longer an option (or afterthought) for field devices but a key requirement that needs to be monitored and evaluated during all phases of the development life cycle. Some of the possible options that can be considered while designing for optimum power consumption are provided in the following list:

- **Selection of optimum connectivity protocols**: Connectivity with the backend server is one of the major factors draining battery power. Fortunately, there are multiple connectivity protocols that are available, and each offers a choice of power consumption along with other trade-offs (such as supported bandwidth, range, and so on). The *IoT connectivity protocols* section later in the chapter provides a list of connectivity protocols along with other parameters that can be used to select the right-fit connectivity protocol as per the available power budget while accommodating other requirements such as coverage, bandwidth, price sensitivity, and security.

- **Relegating the (heavy) processing responsibility to the central server**: Relegating the complex processing (video analytics, complex AI/ML processing, and so on) to the central server rather than performing it at the edge (local processing) is another option for conserving the power of field devices. However, there is a trade-off involved, as more power (and bandwidth) would then be required for data transmission (sending input data to the central server as well as receiving processed data). The volume of data transmitted can be optimized by using compression techniques; however, it would again result in increased power consumption.

- **Reducing data transmission overheads**: One technique (especially relevant for non-time sensitive workloads) is to club multiple data transmission packets into a single payload/packet. This is effective as both connectivity establishment and connectivity severance are expensive operations from a power consumption standpoint. Similarly, sending **Time to Live (TTL)** guidance/instruction along with a timestamp in the payload helps the device to ignore stale messages and is another common technique for conserving power. Sending data on the trigger of an event (for example, a change of sensor value) rather than sending data at a set periodicity is another mechanism that can help in conserving both transmission bandwidth as well as power consumption.

- **Optimizing the amount of data transferred**: Optimizing the amount of data transferred would also have a direct bearing on the power consumption. The simplest example would be sending a mnemonic to the central server (*C, F*) instead of full-length strings (*Centigrade, Fahrenheit*). Understanding the level of precision required for a particular use case can help optimize the volume of data sent to the central server. For example, a smart home use case seldom requires temperature/humidity data to be sent to the central server with more than one digit after the decimal point.

- **Leveraging multiple power modes**: Implementing hardware sleep cycles during idle times is also a common strategy for conserving the power of the battery-operated field devices. Most IoT devices provide three power modes that can be effectively used for minimizing battery consumption:

  - **Normal/no sleep mode**: This mode consumes maximum power and is used in case full functionality (or maximum performance) is needed

  - **Light sleep**: This mode consumes less power than normal mode and is normally achieved by suspending the operation of the processor/MCU and internal clock

  - **Deep sleep**: This mode consumes the least amount of battery power as all the components other than the **real-time clock (RTC)** are in a suspended (hibernation) state

- **Selection of right-fit hardware components**: Selecting hardware, primarily the processor or **microcontroller unit (MCU)**, is a major factor that governs power consumption. Hence, due diligence should be exercised while selecting processors for field devices. Processors that consume less energy and support low-power/energy-efficient modes can provide direct power savings and should be preferred.

- **Source code optimizations**: General source code optimizations (such as the optimization of database queries), minimizing the number of data/status polling operations (operations that are interrupt-driven are normally more efficient than polling operations), configuration-based enabling/disabling of features/functionalities, and other generic techniques such as minimizing the frequency of firmware updates ensure optimum power consumption and longer battery life.

- **Rigorous evaluation of battery performance**: Rigorous testing of the battery during development or QA cycles under different loads or ambient conditions and determining the battery life/efficiency after repeated charge and discharge cycles provides a good prediction of the battery's performance in the field.

- **Perpetual energy availability using energy harvesting techniques**: As mentioned in *Chapter 12*, energy harvesting (energy derived from vibration, solar, motion, and so on) techniques can provide perpetual battery life. Often, energy harvesting is used to augment the power budget by supplementing the power from a normal battery.

It is evident that multiple approaches that include both theoretical (understanding the vendor's data sheet to determine the battery suitability) as well as practical (laboratory experiments) aspects are needed to find the right balance between power consumption and other requirements (for example, processing speed). In fact, it is quite possible that data sheets may not point out the limitations and challenges that the battery may encounter while deployed in the field. Hence, it is important to pose the right set of questions to the vendor about the different conditions under which the battery would operate. The battery vendor should ideally replicate the field environmental conditions within the lab and should be willing to share the test reports and/or related compliance certifications.

Cost optimization

The best **cost optimization** technique is the one that costs nothing; before embarking on the implementation of any IoT project, it is critical to determine whether the envisaged solution would really solve the business problem or would provide the envisaged business benefits. Having the right balance of technical and business minds in the team would certainly help to flag the cases with low to no business value rather than waiting for market acceptability, thus saving crucial investments.

Similarly, having a business mindset (along with technical know-how) would also help teams in other crucial decisions. As an example, it would help teams to make correct *build versus buy* decisions by considering all the technical/commercial implications (in some cases, it is prudent to integrate readymade components available on the market rather than developing them in-house).

One way of looking at IoT use cases is implementing **process automation** solutions, whereby some process element or workflow that was previously handled/performed by a human is augmented or enhanced by a combination of technologies spanning the digital and physical worlds. As a result, most of the considerations that are applicable to process automation solutions/projects are applicable to IoT projects as well – for example, first documenting the existing workflows in the form of process flows for a better understanding of existing flows and then marking parts of the workflow that can be automated using IoT technologies. Documented workflows should also clearly highlight the process elements that would be performed manually even after solution deployment (such as battery replacement).

An example (from a purely commercial standpoint) would be to compare the cost of an operation currently performed by a human and the cost of the same operation (in the long term) if it is planned to be replaced by IoT and related technologies (process automation). An important point is to consider the fact that the cost of human effort varies from one geography to another, indicating that the automation of the same operation may make commercial sense in one geography but may not make financial sense in another geography.

Cost optimization opportunities are scattered during each phase of solution development and deployment. As an example, if a solution is deployed on the cloud, architects need to keep an eye on all the new services/functionalities that cloud providers are launching on a continuous basis. It is quite possible that replacing custom-built service/logic can be swapped for a cloud provider's equivalent service, which simplifies the overall architecture while, at the same time, providing operational cost benefits.

This concludes this section where we have covered two representative NFRs (cost and battery optimization) in detail. The next section covers another important topic, connectivity protocols, and will provide a view of the connectivity options that are available at different layers of the network stack along with their specific usage scenarios.

IoT connectivity protocols

When finalizing the architecture for IoT solutions, there are multiple protocol options (at each layer of the network/connectivity stack) to choose from, as shown in the following figure:

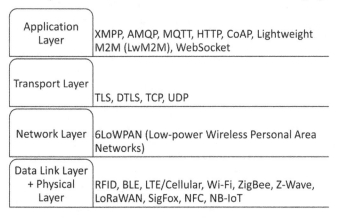

Figure 13.4 – IoT connectivity protocols mapped to different layers of the networking stack

> **Important note**
> The preceding figure provides a representative list of commonly used protocols and in no way should be considered an exhaustive listing.

The multiplicity of connectivity protocols at each layer does result in some amount of ambiguity or perplexity. However, this diverse set of protocols also provides us with a certain degree of flexibility in choosing the protocol that best matches the specific solution requirements such as bandwidth available, allowed range, power consumption, and so on. In general, protocols at each layer can be distinguished based on one or more of the following listed parameters:

- Number of **Quality-of-Service (QoS)** levels supported
- Type of topology supported (mesh, star, point-to-point, bus, tree, and so on)
- Data transfer rate
- Supported frequency bands
- Cost (upfront and recurring)
- Power requirements
- Proprietary/open source
- Range
- Messaging pattern (publish-subscribe, request-response)
- Type and level of security provided
- Compute and memory footprint requirements

There is extensive documentation available in the public domain for all the protocols listed in *Figure 13.4* and it wouldn't be prudent to replicate the same information in this section. Detailed comparisons of protocols based on criteria such as bandwidth and power consumption are also similarly available in the public domain. Before we summarize these, please refer to the hierarchical nature of the layers in the network stack, as shown in *Figure 13.4*. Most of the decisions related to connectivity while architecting IoT solutions relate to the application and data link/physical layer. Accordingly, the options for these two layers are explained in detail in this section.

Now, let's take a look at the applicability considerations for commonly used protocols at the **application layer**:

Connectivity Protocol	Applicability considerations
Constrained Application Protocol (CoAP)	• Lightweight UDP-based protocol, especially suitable for battery-constrained devices as well as for networks with intermittent connectivity • Some similarities to HTTP (for example, methods such as GET, PUT, DELETE, and POST), resulting in a shorter learning curve • Not suitable for cases where delivery confirmation is needed • Typical use cases: smart metering, smart building, and smart energy

Connectivity Protocol	Applicability considerations
Message Queuing Telemetry Transport (MQTT)	• Lightweight protocol, especially suitable for battery-constrained devices • Essentially a broker that helps to decouple data ingestion (publishers) from data processing (subscribers) • Provides scalability and consumes very little bandwidth (non-chatty) • Low footprint, making the protocol suitable for memory-constraint devices
Extensible Messaging and Presence Protocol (XMPP)	• Allows for real-time data exchange of structured data between two nodes • Not suitable for constraint networks/devices owing to its chatty nature
Advanced Message Queuing Protocol (AMQP)	• Not suitable for constraint devices but can be used with systems with strong compute/memory and mains supply • Reliability and interoperability are key benefits • The asynchronous nature (using queuing mechanism) helps to buffer traffic spikes as well as handle intermittent connectivity issues
Lightweight Machine to Machine (LwM2M)	• Built over CoAP and designed specifically for resource-constrained devices • Developed specifically to support remote management of field devices as well as providing strong support for device telemetry • Lacks widespread adoption
Hyper Text Transfer Protocol (HTTP)	• Chatty protocol; not suitable for resource-constraint devices as it consumes a lot of bandwidth as well as power • The requirement of synchronous connection establishment makes this protocol unsuitable for IoT use cases, especially the ones where field devices face intermittent connectivity • No learning curve is involved, as skills developed for web applications can be leveraged
WebSocket	• Capable of sending the data in both directions (device to central server, and vice versa) • Allows for data to be transmitted with minimal latency, ideal for real-time visualizations

Table 13.1 – Applicability considerations for application-layer connectivity protocols

After understanding the pros and cons of common protocols at the application layer, let us focus on the **data link/physical layer** and understand the suitability of protocols at this layer under different operating conditions/environments:

Connectivity Protocol	Applicability considerations
Wi-Fi	• Ubiquitous nature • Consumes high power and is generally preferred for devices with mains supply • Supports high bandwidth and is suitable for streaming videos, audio, large datasets, and so on
ZigBee	• Useful in cases where transmission requires low power and low data rates • Self-configuring and self-healing (mesh topology allows faulty nodes to be bypassed for uninterrupted communication) properties result in resilient architectures • Supports multiple network topologies – mesh, multi-point, and point-to-point • Node-to-node communication range is smaller; however, can support long distances using a mesh topology, which also provides high scalability • Typical use cases: home automation and building management systems
Z-Wave	• Proprietary standard providing capabilities of self-configuration and self-healing (like ZigBee) • Provides ultra-low latency • Typical use cases: home automation
Bluetooth Low Energy (BLE)	• Extremely power-efficient and so suitable for power or battery-constrained devices • Used for connecting wireless electronic devices (keyboard, mouse, headset, printer, external speakers, and so on) • Is generally used in cases where cell phones can be used as DG as cell phones invariably support Bluetooth • Not suitable for sending large amounts of data and over large distances • Supports both point-to-point as well as mesh topology • Typical use cases: fitness trackers, smart wearables, and so on

Cellular, Long-term evolution (LTE)	• Provides low latency, good coverage, high bandwidth, and long range
	• Protocol consumes considerable power so may not be suitable for power-sensitive devices
	• Better at penetrating closes spaces such as indoor walls
	• Involves a recurring charge
Narrow Band-Internet of Things (NB-IoT)	• Cellular network specially designed for IoT use cases
	• Draws very little power; however, coverage is limited
	• Suitable for stationary field devices
Near-Field Communication (NFC)	• Allows for communication between devices that are in proximity (distance not more than 4 cm)
Radio Frequency Identification (RFID)	• Encodes a unique value into a tag or label
	• Similar to bar codes (or other paper-based tags) but is much more robust and doesn't require line of sight for detection
	• Multiple tags can be read at a time
	• Typical use cases: asset tracking and automatic vehicle toll collection
Long-Range Wide Area Network (LoRaWAN)	• Consumes very little power and can support a large number of field devices (~50,000 per cell), making it useful in high-density deployments (such as large cities)
	• Reasonably good coverage and supports both outdoor as well as indoor operation
	• Has limited bandwidth making firmware updates difficult; suitable primarily for low data rate applications
	• Typical use cases: smart city applications such as smart lighting, smart garbage collection, and so on

Table 13.2 – Applicability considerations for data link/physical layer connectivity protocols

This brief discussion on the connectivity protocols was the last section in this chapter so now, let's summarize what we have learned in this chapter.

Summary

In this final chapter, we covered practical issues and challenges that are typically encountered while implementing IoT solutions/use cases. It listed key lessons learned from my past IoT projects. Specific references were made to two frameworks (Bain's and Porter's) as these can help to gauge as well as articulate the benefits/value of the envisaged IoT solution(s). It also listed the key NFRs and emphasized the importance of focusing on them during the complete project life cycle and covered two examples (power and cost optimization) in detail. Finally, it provided a representative list of IoT connectivity protocols, along with the scenarios in which they are most appropriate.

The IoT ecosystem, in general, is very dynamic, new techniques/technologies are continuously being discovered, and existing ones getting refined/evolved. Going forward, the trend is expected to continue with advancements expected at all layers of the IoT reference architecture. Some of the future possibilities are as follows:

- Sensors or field devices continuing their miniaturization journey with smaller batteries leading the pack.

- There would be more focus on making the field devices operate perpetually by leveraging energy harvesting techniques.

- With stringent sustainability requirements, ensuring that the entire solution (field devices, central server, and other connectivity devices such as routers and switches) is energy-efficient would be a mandate rather than an option.

- New connectivity/network technologies would keep on pushing the boundaries of network bandwidth, device density, and latency, providing more flexibility to solution designers for implementing processing logic at the edge or central server, or a combination of both.

- Enhanced compute capabilities both at the edge/DG (improved microprocessors/microcontrollers) as well as at the central server (quantum technology) would provide grounds for developing more powerful and innovative compute-intensive solutions.

The book started with the premise of providing guidance for developing innovative solutions using IoT technology and capabilities. I sincerely hope that prior chapters and this chapter were able to fulfill that objective in a concise yet effective manner.

Finally, I would like to thank you for spending your time, energy, and money on this book. Wishing you all the success on your journey of implementing innovative IoT solutions!

Index

Symbols

A

www.packtpub.com

Subscribe to our online digital library for full access to over 7,000 books and videos, as well as industry leading tools to help you plan your personal development and advance your career. For more information, please visit our website.

Why subscribe?

- Spend less time learning and more time coding with practical eBooks and Videos from over 4,000 industry professionals

- Improve your learning with Skill Plans built especially for you

- Get a free eBook or video every month

- Fully searchable for easy access to vital information

- Copy and paste, print, and bookmark content

Did you know that Packt offers eBook versions of every book published, with PDF and ePub files available? You can upgrade to the eBook version at packtpub.com and as a print book customer, you are entitled to a discount on the eBook copy. Get in touch with us at customercare@packtpub.com for more details.

At www.packtpub.com, you can also read a collection of free technical articles, sign up for a range of free newsletters, and receive exclusive discounts and offers on Packt books and eBooks.

Other Books You May Enjoy

If you enjoyed this book, you may be interested in these other books by Packt:

Implementing Cellular IoT Solutions for Digital Transformation

Dennis McCain

ISBN: 978-1-80461-615-4

- Understand how IoT enables an enterprise's digital transformation
- Discover the applications of various IoT wireless technologies
- Explore IoT devices, architectures, and real-world use cases
- Dive deep into LTE and 5G cellular technologies and how they enable IoT
- Build a privacy and security framework in an IoT solution
- Select the best components for a cellular IoT enterprise solution
- Overcome challenges in the IoT solution life cycle
- Examine new cellular IoT technologies, trends, and business models

Designing Production-Grade and Large-Scale IoT Solutions

Mohamed Abdelaziz

ISBN: 978-1-83882-925-4

- Understand the detailed anatomy of IoT solutions and explore their building blocks
- Explore IoT connectivity options and protocols used in designing IoT solutions
- Understand the value of IoT platforms in building IoT solutions
- Explore real-time operating systems used in microcontrollers
- Automate device administration tasks with IoT device management
- Master different architecture paradigms and decisions in IoT solutions
- Build and gain insights from IoT analytics solutions
- Get an overview of IoT solution operational excellence pillars

Packt is searching for authors like you

If you're interested in becoming an author for Packt, please visit `authors.packtpub.com` and apply today. We have worked with thousands of developers and tech professionals, just like you, to help them share their insight with the global tech community. You can make a general application, apply for a specific hot topic that we are recruiting an author for, or submit your own idea.

Share Your Thoughts

Now you've finished *Architectural Patterns and Techniques for Developing IoT Solutions*, we'd love to hear your thoughts! Scan the QR code below to go straight to the Amazon review page for this book and share your feedback or leave a review on the site that you purchased it from.

`https://packt.link/r/1803245492`

Your review is important to us and the tech community and will help us make sure we're delivering excellent quality content.

Download a free PDF copy of this book

Thanks for purchasing this book!

Do you like to read on the go but are unable to carry your print books everywhere?

Is your eBook purchase not compatible with the device of your choice?

Don't worry, now with every Packt book you get a DRM-free PDF version of that book at no cost.

Read anywhere, any place, on any device. Search, copy, and paste code from your favorite technical books directly into your application.

The perks don't stop there, you can get exclusive access to discounts, newsletters, and great free content in your inbox daily

Follow these simple steps to get the benefits:

1. Scan the QR code or visit the link below

https://packt.link/free-ebook/9781803245492

2. Submit your proof of purchase
3. That's it! We'll send your free PDF and other benefits to your email directly

www.ingramcontent.com/pod-product-compliance
Lightning Source LLC
Chambersburg PA
CBHW080627060326
40690CB00021B/4841